SOYBEAN DISEASES
of the North Central Region

Edited by
T. D. Wyllie and D. H. Scott

APS PRESS
The American Phytopathological Society
St. Paul, Minnesota

Financial Contributors

Asgrow Seed Company, subsidiary of Upjohn Company
CIBA-GEIGY Corporation, Agricultural Division
Cooperative State Research Service,
U.S. Department of Agriculture
DeKalb-Pfizer Genetics
Funk Seeds International
Gustafson, Inc.
ICI Americas, Inc., Biological Research Center
Monsanto Agricultural Products Co.
Sandoz Crop Protection Corporation

This book has been reproduced directly
from typewritten copy submitted in final form
to APS Press by the editors of the volume.
No editing or proofreading has been done by the Press.

Library of Congress Catalog Card Number: 88-70266
International Standard Book Number: 0-89054-087-X

Printed in the United States of America

The American Phytopathological Society
3340 Pilot Knob Road
St. Paul, Minnesota 55121, USA

TABLE OF CONTENTS

PREFACE

The purpose of this proceedings is to provide the latest information available on the major diseases affecting soybeans in the north central soybean growing area of the United States. Hopefully, this compilation of research information will serve as a useful reference source for plant pathologists, agronomists, plant breeders, extension specialists, crop consultants, state and federal agencies and others interested in the development, growth and production of soybeans. The bibliography contained in this proceedings may well be the most current and extensive reference source on certain soybean diseases that is available at this time.

These proceedings are the direct result of a North Central Region Soybean Disease Workshop held in Indianapolis, Indiana on March 10 and 11, 1987. This workshop was sponsored by the North Central Regional Committee on soybean diseases (NCR-137). The NCR-137 committee is composed of representatives from the Agricultural Experiment Stations in Illinois, Indiana, Iowa, Kansas, Michigan, Minnesota, Missouri, Nebraska, North Dakota, Ohio, South Dakota and Wisconsin.

The editors are indebted to each of the authors who took time from their busy schedules to prepare their presentations in manuscript form suitable for publication. Their preparations made our tasks much easier. We are also appreciative of the support and efforts provided by each of the members of the NCR-137 committee, and to Dr. Robert G. Gast Administrative Advisor to NCR-137. The editors are also appreciative of the effort of Anita Blanchar who typed and formatted the manuscript. A brief professional sketch of each author follows.

Dr. John M. Barnes, Principal Scientist, CSRS, Plant and Animal Studies, USDA, Washington, D.C. has served with the USDA since 1967. His duties include the evaluation, planning and funding of research projects in the plant sciences and particularly in plant pathology and nematology in cooperation with the land grant universities in the United States. He is currently developing research information networks, preparing budget profiles for plant pathology and nematology and making a study of research funding changes and their effects on the buying power of research funds. He also serves as an extra-curricular co-editor of the Applied Botany and Crop Science Series, Academic Press, London. Dr. Barnes received his Ph.D. degree from Cornell University, Ithaca, New York in 1960.

Dr. Richard S. Ferriss, Associate Professor of Plant Pathology, Department of Plant Pathology, University of Kentucky, Lexington, Kentucky has been heavily involved in soybean pathology research since joining the faculty in 1979. His research includes the epidemiology of seedling diseases and the relationship of seedling diseases to seed quality in soybeans. Dr. Ferriss has several important research publications in these areas of research. He received his Ph.D. degree from the University of Florida, Gainesville, Florida in 1979.

Mr. Ademir Henning is currently working towards his Ph.D. degree under the direction of Dr. Denis C. McGee at Iowa State University, Ames, Iowa. His research concerns soybean seed encapsulation. Prior to studying at Iowa State, Mr. Henning worked for ten years in seed pathology and seed physiology for EMBRAPA, National Center for Soybean Research, Parana State, Brazil.

Dr. Denis C. McGee is a Professor of Plant Pathology and jointly serves the Seed Science Center and the Department of Plant Pathology, Iowa State University, Ames, Iowa. Dr. McGee is a widely recognized authority in seed pathology and seed technology. He is in charge of the seed pathology program for Iowa State, and his current research includes the pathology of soybean and other crop seeds. He has over forty research publications and a like number of extension publications. He has written book chapters dealing with soybean seed treatment and the pathological interactions that affect seed deterioration, and he has just finished a chapter on maize diseases for a seed technologists reference source. Dr. McGee received his Ph.D. from the University of Edinburg, Scotland in 1967. He served for a period with the University of Maine prior to joining the faculty at Iowa State where he has been for the past nine years.

Dr. John H. Hill, Professor of Plant Pathology, has a joint appointment with the Department of Plant Pathology and the Department of Microbiology at Iowa State University, Ames, Iowa. Dr. Hill is a plant virologist specializing in the application of biotechnology to theoretical and practical problems in plant virology. His research includes the use of such technologies as cloning, DNA sequencing and the preparation and use of monoclonal antibodies to detect viruses in soybeans and other crops. He has authored or co-authored over fifty research publications and has contributed book chapters on methods in enzymatic analyses and the detection of soybean mosaic virus in soybean seeds. Dr. Hill has been at Iowa State University since receiving his Ph.D. degree from the University of California, Davis, California in 1971.

Dr. David B. Sauer, Research Plant Pathologist, USDA, ARS, U.S. Grain Marketing Research Laboratory, Manhattan, Kansas has an adjunct appointment in the Department of Plant Pathology, Kansas State University, Manhattan, Kansas. His research concerns the biology and control of post harvest fungi in stored grain. Dr. Sauer has published over fifty research papers and has contributed book chapters concerning the microflora of stored grain and on mycotoxins associated with stored grain. Dr. Sauer received his Ph.D. degree from the University of Minnesota, St. Paul, Minnesota in 1967 and has been associated with the USDA since that time.

Dr. T. Scott Abney is a Research Plant Pathologist, USDA, ARS, and Associate Professor of Plant Pathology in the Department of Botany and Plant Pathology, Purdue University, West Lafayette, Indiana. His research centers around the mycological and epidemiological factors affecting soybean diseases with special emphasis on soybean seed quality. Dr. Abney has participated in the development and release of thirteen USDA/Purdue soybean varieties with improved yield and disease resistance. He also has an active role in testing advanced breeding lines and identifying disease resistant germplasm as part of the Northern States Soybean Improvement Program. Dr. Abney has authored or co-authored fifty-three research publications and over forty other publications. He has been associated with the USDA and Purdue University since he received his Ph.D. from Iowa State University, Ames, Iowa in 1967.

Dr. L. Daniel Ploper had just finished the requirements for his Ph.D. degree under the direction of Dr. T. Scott Abney, Department of Botany and Plant Pathology, Purdue University, West Lafayette, Indiana at the time this paper was presented. His research concerned the Diaporthe/Phomopsis complex in soybeans. He currently is Research Plant Pathologist, Tucuman State Agricultural Research Station, Tucuman, Argentina.

Dr. Bill W. Kennedy is a Professor of Plant Pathology in the Department of Plant Pathology, University of Minnesota, St. Paul, Minnesota. Dr. Kennedy was one of the first researchers in Minnesota to work on soybean diseases and has been instrumental in the development of soybeans in that state. His current research deals with the cronopathology of diseased plants; the relationships of timing, plant rhythms and disease development. Dr. Kennedy has authored or co-authored over one hundred forty research publications and six book chapters. He recently was a member of a research tour sponsored by the American Soybean Association and Imperial Chemicals of England. He received his Ph.D. degree from the University of Minnesota, St. Paul, Minnesota in 1961, and joined the faculty the next year.

Dr. John M. Dunleavy, Research Plant Pathologist, USDA, ARS at Iowa State University, Ames, Iowa, is one of the pioneers in the intensive research on fungal diseases of soybeans. He says that he can remember when all the federal and state soybean workers in the United States could hold a joint meeting "Around a ping pong table." His work also includes research on the soybean mosaic virus, and most recently, the bacterial diseases of soybeans. He has over eighty research publications, three book chapters and numerous extension articles to his credit. Dr. Dunleavy has served the USDA since 1953, when he received his Ph.D. degree from the University of Nebraska, Lincoln, Nebraska.

Dr. Paul A. Backman, Professor of Plant Pathology, Department of Plant Pathology, Auburn University, Auburn, Alabama has been heavily involved in soybean and peanut pathology research since joining the faculty in 1972. Dr. Backman's research includes chemical and biological control, disease loss estimates, the effects of rotation and soil amendments on disease severity, and the role of beneficial organisms on leaf surfaces in disease development. He has authored or co-authored eighty research publications, five book chapters and is currently participating in the writing of the third edition of the "Compendium of Soybean Diseases." Dr. Backman received his Ph.D. degree from the University of California, Davis, California in 1970. From that time until he joined the faculty at Auburn University, Dr. Backman had a short post doctoral position at North Carolina State University and then served with the USDA at Clemson, South Carolina.

Dr. Craig R. Grau is a Professor of Plant Pathology in the Department of Plant Pathology, University of Wisconsin, Madison, Wisconsin. Dr. Grau's research centers around the soilborne diseases of legumes, and includes the cultural and host resistance aspects of disease control. His research has made significant contributions to the understanding of the disease known as Sclerotinia stem rot of soybeans. Dr. Grau has authored or co-authored over forty research publications and over sixty extension publications. In addition, he has written a book chapter on powdery mildew of soybeans and is writing the section on Sclerotinia for the third edition of the "Compendium of Soybean Diseases." Dr. Grau received his Ph.D. degree from the University of Minnesota, St. Paul, Minnesota in 1975, and after a period of post doctoral research at North Carolina State University, joined the faculty of the University of Wisconsin in 1976.

Dr. Donald H. Scott is a Professor of Plant Pathology and Extension Plant Pathologist in the Department of Botany and Plant Pathology, Purdue University, West Lafayette, Indiana. His research involves the diseases and disease control of soybeans and other crops, including the effects of tillage practices and crop rotation on disease development. He has authored or co-authored nine research publications, fifty research reports and over one hundred extension publications. Dr. Scott joined the faculty at Purdue University after receiving his Ph.D. degree form the University of Illinois, Urbana, Illinois in 1968.

Dr. A. F. "Fritz" Schmitthenner is a Professor of Plant Pathology in the Department of Plant Pathology, The Ohio State University, OARDC, Wooster, Ohio, "Fritz," as he is known to all his friends, has spent his entire career of some thirty-five years at Ohio State, and is the recognized authority on Phytophthora root rot of soybeans. His research involves the root and stem diseases of soybeans and their control, as well as the ecology of the Pythium and Phytophthora pathogens on all crops. He has authored or co-authored over seventy research publications and thirteen book chapters. Dr. Schmitthenner was honored by the American Phytopathological Society for his significant contributions to the science of plant pathology when he was made an APS Fellow in 1984. He received his Ph.D. degree from The Ohio State University, Columbus, Ohio in 1953.

Elisa F. Smith is a research associate in the Department of Plant Pathology, Auburn University. Miss Smith received her M.S. degree in 1988 on the epidemiology of *Diaporthe phaseolorum* var. *caulivora* in southern soybean. She has been associated with Dr. Backman since 1984 working on peanut and soybean disease problems. She has authored or co-authored several abstracts during her young career and currently has several publications in process.

Dr. Dale I. Edwards, Research Plant Pathologist, USDA, ARS is also an adjunct Associate Professor of Plant Pathology, Department of Plant Pathology, University of Illinois, Urbana, Illinois. Dr. Edwards was one of the first nematologists to do intensive research on the soybean cyst nematode in the North Central Region. His research includes the biology, ecology, host-parasite relations and control of soybean nematodes through rotation, biological control, resistant cultivars and nematicides. He has authored or co-authored over seventy research publications and has written four book chapters concerning nematodes, nematode-fungus interactions and control of nematodes. After serving as a nematologist for a commercial company in Honduras for three years, Dr. Edwards joined the USDA staff in Urbana, Illinois in 1965. He received his Ph.D. degree from the University of Illinois, Urbana in 1962.

Dr. Terry Niblack is an Assistant Professor in the Department of Plant Pathology, University of Missouri, Columbia, Missouri. Her research in nematology has dealt with nematode problems concerning ecological relationships with shade trees, ornamentals, and soybeans. In Georgia, where she received her doctorate, she concentrated on research pertaining to the soybean cyst nematode and the root knot nematode. Recently, while on a post doctoral fellowship at Iowa State University, she has worked in the area of population dynamics, ecological relationships, and host-parasite interactions of the soybean and soybean cyst nematode. Dr. Niblack has approximately twenty research publications. Dr. Niblack received her Ph.D. from the University of Georgia, Athens, Georgia in 1985.

Dr. John L. Lockwood, Professor of Plant Pathology, Department of Botany and Plant Pathology, Michigan State University, East Lansing, Michigan is world renowned for his research on the ecology of root-infecting fungi in the soil. In addition, his research includes soybean root diseases and the biological control of plant diseases. He has authored or co-authored over one hundred-twenty research publications and has written numerous review articles and book chapters. Dr. Lockwood was honored for his significant contributions to the science of plant pathology by the American Phytopathological Society when he was made an APS Fellow in 1977, and he later served as President of that society in 1984-1985. Dr. Lockwood began, and has continued, his illustrious career at Michigan State University, when he joined the faculty in 1955. He received his Ph.D. degree in 1953 from the University of Wisconsin, Madison, Wisconsin.

Dr. James B. Sinclair is a Professor of Plant Pathology in the Department of Plant Pathology, University of Illinois, Urbana, Illinois. He has had many years experience in research on diseases of soybeans, with particular emphasis in seedborne pathogens. In addition, he is heavily involved with international plant pathology programs and has wide experience as an advisor to scientists in foreign countries on soybean diseases, and has guided the research of numerous foreign students. Dr. Sinclair has authored or co-authored over six hundred research articles, research reports, and extension publications. He was editor of two editions of the "Compendium of Soybean Diseases" and is working on the third edition, and has co-authored three books. Dr. Sinclair joined the faculty at the University of Illinois in 1968 after serving twelve years on the faculty of Louisiana State University. He received his Ph.D. degree from the University of Wisconsin in 1955.

Dr. Hideo Tachibana, Research Plant Pathologist, USDA, ARS, and adjunct Professor of Plant Pathology in the Department of Plant Pathology, Iowa State University, Ames, Iowa has spent his entire professional career at this location. He has concentrated most of his research work on brown stem rot of soybeans, on which he is the recognized authority. He has doggedly pursued the development of soybean varieties resistant to this disease and the concept of "prescribed resistant varieties," which advocates their use only when needed for brown stem rot control. He has authored or co-authored over sixty-five research publications and has written a book chapter on the bacterial diseases of soybeans. Dr. Tachibana received his Ph.D. degree from Washington State University, Pullman, Washington in 1963.

Dr. Thomas D. Wyllie is a Professor of Plant Pathology in the Department of Plant Pathology, University of Missouri, Columbia, Missouri. His research for the past twenty-seven years has been in the area of soilborne fungal pathogens, and especially on the difficult to study disease known as charcoal root rot. The research includes evaluating germplasm for resistance, and more recently a comprehensive field study of factors affecting populations of the pathogen in the soil with the long term goal of forecasting disease probability from soil and other parameters. Dr. Wyllie has over ninety-five research publications and was editor for a three volume series of books on mycotoxins. Dr. Wyllie joined the University of Missouri faculty after receiving his Ph.D. degree from the University of Minnesota, St. Paul, Minnesota in 1960.

Donald H. Scott
Thomas D. Wyllie
Editors

CHALLENGES FOR INTERDISCIPLINARY AGRICULTURAL RESEARCH

John M. Barnes

USDA-CSRS
Washington, DC 20251

Research programs aimed at increasing yields and improving quality of U.S. crops have clearly been effective in providing new knowledge as a basis for the outstanding productivity of our system. However, we are faced with the mixed blessing of domestic surpluses coupled with international trade imbalances and the resultant depressed farm economy. Debt reduction directives are forcing major realignments in our agricultural research funding processes and raising concerns over our ability to maintain a healthy research base to meet unknown future emergencies. Thus, we are forced to seek better ways to invest static or declining funding bases in critical areas of need. Discipline strength has been a hallmark of progress in U.S. agricultural technology. U.S. agriculture has entered the high technology arena of increasing complexity. This is mirrored in the fast changing disciplines of science, most of which are composed of several subdisciplines.

A traditional strength of each discipline has been their depth of specialization and closely protected autonomy. Strong departments attract top students; strong departments attract top faculty; top faculty attract the best funding; the best funding strengthens the research and teaching programs of departments, which attracts more top students and faculty--and the cycle is self-reinforcing. This is good. However, these positions of strength usually do not encourage an open dialogue on the changes that immediate pressures dictate to us.

Commodity development programs in the universities usually consist of "cooperative" efforts wherein the cooperation is often loose, at best. In programs where there is close, continuous cooperation, it is generally because of individual scientist initiative and not because of institutional or funding source policy. In industry, research and development teams are built around the product or major goal, with the team being the unit for cost accountability and productivity. This kind of system is anathema to academe. When individual scientists chart their own courses in commodity-oriented research and development projects, conferring occasionally, we find too many instances in which a major step, such as germplasm or variety release, must be delayed (or even aborted) because of some late-breaking flaw detected by a scientist working independently.

In view of the stresses just mentioned, it is becoming evident to many research leaders that cross-discipline collaboration bridging discipline strengths is an important approach to be used in certain research and development projects. For example, a deep commitment by soybean breeders and geneticists, pathologists, entomologists, soil scientists, economists, etc., in planning and executing soybean variety improvement programs in the universities could, I believe, result in greater savings in time and funds in producing desired results.

The "New Biotechnology" initiative, through excellent cooperation between the Science and Education component of USDA and the land-grant university network, has forged an unparalleled opportunity for the agricultural sector to capitalize on a lead role in providing high technology products undreamed of one or two decades ago. Now, this initiative has caused many universities to restructure research organization, provide new funding mechanisms, build new facilities, and develop new arrangements with the biotechnology industry. Plant pathologists and those in kindred sciences working on soybeans have knowledge bases through which we can exploit future technological breakthroughs. Moreover, we need to make sure that science administrators appreciate the crucial roles of those who work with the host plant and organisms affecting the health of our crops. At the same time, colleagues have argued that we must not neglect practical information and knowledge; we must also train and educate people who know how to grow plants, how to identify them and how to identify the organisms that attack them.

Undergraduate enrollments in Colleges of Agriculture continue a general decline, as prospects for other careers lure incoming students. A recent survey conducted by USDA in cooperation with Texas A and M University reports that over the next five years 48,000 positions will be available in employing graduates of agriculture and veterinary colleges, and only 44,000 graduates will be available, amounting to a ten percent shortfall. Scientists, engineers, managers, sales representatives, and marketing specialists will account for three-fourths of the total employment openings for new college graduates. Today is still an exciting time to be involved in the food and agricultural enterprises of our nation. Our professional science societies, universities, federal and private research organizations, commodity associations, and others should play much more active roles in public relations, working together to familiarize students at all levels of education of the opportunities.

Today, there is much discussion about the contributions of the Integrated Pest Management (IPM) concept. There have been many accomplishments from IPM over the almost two decades since the national program began. However, there are many science leaders who feel we must build outward, reflecting a more holistic approach to the realities of the farm-forestry enterprises. We look forward to finding a way to attain a comprehensive dialogue among all the scientists having roles in this building process. This means that the pest sciences and the production sciences would join with the social sciences as equal partners in a national enterprise to shape a strategy and program built on a solid, long-term view of what the U.S. agricultural enterprise would look like and would require in ten to twenty years.

In summary, I have outlined some areas that I think are crucial to reestablishing the health of the U.S. agricultural science and education system. These challenges must be viewed in a positive way, as we consider the new agricultural technologies and products on the horizon. Some colleagues have argued that we cannot predict what the needs will be in five years, let alone one or two decades. My answer is that no one else can do it better than we, so let's take the lead and actively think about the future, rather than continuing to react to events as they happen.

SEED DISEASES

T. S. Abney and L. D. Ploper

USDA/ARS and Department of Botany and Plant Pathology
Purdue University
West Lafayette, Indiana

The soybean [*Glycine max* (L.) Merrill] is one of the most important field crops because of its many food, feed, and industrial uses. Although its origin and early history remain obscure, it is generally believed to be a native of eastern Asia. It has been used for thousands of years as an important source of vegetable oil for human consumption and protein for animal feed.

All parts of the soybean plant are susceptible to a number of diseases which reduce the quality and quantity of seed yields. Over 100 pathogens are known to attack this crop; of these, 35 are of economic importance (469). Some of these pathogens may be very destructive one season and difficult or impossible to find the next season, while others are endemic and take a yearly toll on production. Estimated yield losses due to diseases are commonly cited as 9.8% from fungi, 0.1% from bacteria, 0.2% from viruses, and 4.8% from nematodes, for a total loss due to all pathogens of 14.9% (398). More effective disease control is essential if improved quality and quantity of soybean production are to be realized.

Soybean diseases that affect seeds have received recent attention because seedborne organisms reduce the commercial grade of grain at the elevator by causing smaller, distorted, spotted, and stromatized seeds. Infection of developing or mature seeds is important not only because it can cause these direct losses, expressed in terms of yield and quality reduction, but also because it can reduce germination. In addition to producing poor stands and weak seedlings, infected seeds can serve as a means of survival for many pathogens.

The kinds of fungi that infect soybean seeds, the extent to which they do so, and their effect on seeds have received much emphasis recently. The reasons for this concern are related to the recent intensive and extensive cultivation of soybeans which, has exposed them to a greater variety of fungi that are potentially pathogenic to seeds, i.e., able to reduce seed yield, germinability, and/or quality. At least 26 fungal, seven bacterial, and seven virus diseases in soybeans are seedborne. A number of other fungi are associated with soybean seeds (411), but not all of them have been directly implicated in disease.

The most consistently cited fungi isolated from soybean seeds are *Diaporthe*/*Phomopsis* spp., *Cercospora kikuchii*, *Alternaria* spp., and members of the genus *Fusarium* (398,423,469). The most common and endemic seed diseases are purple seed stain, caused by *Cercospora kikuchii* (Matsumoto & Tomoyasu) Gardner, and pod and stem blight and/or Phomopsis seed decay, caused by various species of *Diaporthe* and *Phomopsis* (425). Among the fungi associated with the *Diaporthe*/*Phomopsis* disease complex are: *Diaporthe* /*phaseolorum* (Cke. & Ell.) Sacc. var. *sojae* (Lehman) Wehm. [anamorph: *Phomopsis sojae* Lehman], *Diaporthe phaseolorum* (Che. & Ell.) Sacc. var. *caulivora* Athow and Caldwell, and *Phomopsis longicolla* Hobbs (22,244,449). Various investigators have studied the etiology of soybean diseases caused by members of the genus *Diaporthe*. However, there has been a recent separation of the highly variable pycnidial state into *Phomopsis longicolla* and *Phomopsis sojae* (244). This separation was based on a limited number of isolates and without a definitive assessment of the teleomorphic association for Phomopsis longicolla. Furthermore, there is still controversy about the separation of Diaporthe phaseolorum into taxonomic varieties. This awareness has prompted some researchers to refer to this grouping of organism(s) as Diaporthe/Phomopsis spp. or complex (333).

Seed infection caused by Diaporthe and Phomopsis spp. is the predominant problem observed in most countries where soybeans are grown. Diaporthe/Phomopsis seed infection causes poor seed germination and seedling emergence. In addition, members of the Diaporthe /Phomopsis complex are associated with foliar and stem diseases (469). The stem blight phase associated with Diaporthe phaseolorum var. sojae generally has little or no economic importance compared to stem canker damage associated with D. phaseolorum var. caulivora (22,449). Stem canker was a serious problem only in the North Central states in the early 1950's, but in recent years has caused important losses in the Southern states (28). There is apparent unanimity among soybean research and extension personnel on the economic importance of the seed decay component of the Diaporthe/Phomopsis complex. Infected seeds may appear normal or be shrunken, cracked, visibly moldy, and lighter in weight than healthy seeds (423). Oil and flour derived from infected seeds are of lower quality than that from non-infected seeds (423).

Several control measures have been proposed to reduce the incidence of soybean seed infection by fungal pathogens. These measures include cultural practices (528), the use of late-maturing cultivars (513), chemical treatments (3), and genetic resistance (420). Cultural practices such as crop rotation, tillage, fertilization, and planting date are effective in reducing the incidence of seedborne fungi (425,528). Late-maturing cultivars can escape infection because maturation occurs when environmental conditions (moisture and temperature) are not favorable for seed infection (333,549).

Chemical treatments include fungicide seed treatments and foliar sprays. Improvement of soybean stands and yields from seed treatment of poor quality seed has been reported (90,532). Single or split applications of foliar fungicides have also been effective in reducing, but not eliminating, the incidence of seedborne pathogens. Foliar fungicides have been used only on a limited basis because they must be applied before disease symptoms are visible and before it is known if economically significant levels of seed infection will occur (417,505). The incidence of seedborne pathogens is dependent on both inoculum availability and field environment during seed maturation. Several states and private companies have developed predictive systems to more effectively schedule fungicide application.

Selection of soybean cultivars genetically resistant to these diseases could be one of the best means of crop protection and is generally the most economical and efficient way of controlling diseases. To obtain these cultivars, sources of resistance need to be identified and then resistance incorporated into modern cultivars.

Identification of sources of resistance has been limited because of the strong influence that time of maturity and particularly environmental conditions between the yellow pod and mature pod stages have on seed quality (549). Warm temperatures and high humidity during these final stages of seed maturation are known to enhance disease development. The differences in seed disease associated with varying maturity times are generally explained by these environmental factors rather than by genotypic resistance (513). Cultivars that mature early in the season generally have a higher incidence of seedborne fungi than do cultivars maturing later in the growing season when lower temperatures and moisture are not as conducive to seed infection. Care must be taken to separate environmental and host maturity effects before such resistance can be evaluated.

Improving techniques for measuring the degree of resistance is probably the key factor in determining the success or failure of a program screening for disease resistance. Some breeders use the "delayed harvest" technique (550), but very few breeding programs actually employ inoculation with the most important pathogens as a means of decreasing disease escapes. Similarly, bioassays for measuring actual seed infection, instead of visual observation, are performed infrequently.

Identification of disease resistance is further complicated by the existence of pathogenic variability in these fungi. Although no physiologic races have been defined in C. kikuchii or any of the components of the Diaporthe/Phomopsis complex, it is known that differences in virulence, temperature and media requirements, colony morphology, sporulation ability, etc., occur within these species (126,270,437). The evaluation of pathogenicity and virulence becomes critical when inoculations are part of a screening procedure.

Despite all these limitations, differences in susceptibility to seed diseases have been reported for cultivars and strains of several maturity groups (43,64). Even genotypes growing and maturing under near-identical conditions exhibit great variation in percent of infected seeds (35,398). Although resistance or tolerance to seed diseases has not been characterized genetically, differences among genotypes suggest that some mechanism of limiting fungal invasion of the seeds is present in those genotypes that consistently show a low incidence of infected seeds.

In recent years, researchers have been trying to identify features of the soybean plant (morphological, physiological, etc.) which are associated with resistance or tolerance to seed infection. Growth habit of the soybean plant is one of the characteristics that has been investigated. Soybean genotypes are described as having a determinate or indeterminate growth habit, but intermediate types have also been characterized. Indeterminate genotypes continue to increase in height after they begin flowering and flower over a longer time period. Genotypes with a determinate growth habit have very little increase in height after initiation of flowering. Thomison and Kenworthy (514) speculated that resistance to seed diseases could be linked to growth type. In their work with Clark near-isogenic lines, isolines with indeterminate and semideterminate growth habit exhibited less seed infection and higher germination levels than determinates of similar maturity. However, Balles and Abney (35) found no difference in seed infection of near-isogenic lines of Clark and Harosoy when plants were inoculated with C. kikuchii. Thus, evidence of an association of growth type with seed disease reaction is inconclusive.

Another morphological feature that may influence seed infection is the plant's pubescence morphology (217). The soybean plant possesses a dense covering of erect hairs or trichomes on the epidermal surface of its leaflets, petioles, stems, calices, and pods. Much genetic variation exists with respect to size, shape, color, durability, and density of the trichomes (244,479). The soybean genotype PI 80837, identified as having resistance to seed infection, has dense pubescence (423,425). In another study it was indicated that glabrous types have better seed quality than normal pubescent types (479). These results were based on natural infection and a visual estimation of seed quality.

Another characteristic that has been thought to be directly related to seed infection is "hard-seededness," which gives a type of seed dormancy due to reduced permeability of the seed to water (513). Researchers have found that genotypes with hard, impermeable seed often have less seed infection and higher germination than do genotypes with permeable seed coats (232). The thicker seed coats of the hard seeds may present an effective mechanical barrier to fungal penetration because of the presence of wax/cutin deposits embedded in the seed coat surface and the reduced number of pores present in this surface (563). Fungal entry via these pores has been found to occur without the presence of visible seed coat defects (232). The hard seed characteristic is an undesirable trait of soybeans and breeders have selected against this characteristic because of its adverse effect on germination and stand establishment.

A host characteristic in genotypes that mature in the same environment has been identified which has a direct influence in the development of seed diseases. Results from Indiana studies, involving diverse germplasm sources and growth regulators, suggest that rate of late season maturation dramatically influence seed infection (4,35,398).

SEEDLING ESTABLISHMENT - AN EPIDEMIOLOGICAL PERSPECTIVE

Richard S. Ferriss

Department of Plant Pathology
Agricultural Experiment Station
University of Kentucky
Lexington, KY 40546-0019

A plant can be considered to be established when it is fully able to support its own growth by photosynthesis. This attainment of self-sufficiency results when heterotrophic growth (relying primarily on materials stored in the seed) shifts to autotrophic growth, and is a natural demarcation of the end of the seedling stage (104). For soybean, establishment occurs at about the time that the first true leaves expand (326) and corresponds to stage V_1 (152).

In a broad sense, the epidemiology of soybean seedling disease is concerned with the reasons why viable, planted seeds sometimes fail to produce established plants. There are many potential causes of such failure. Some of these causes are abiotic, but most of them probably involve at least one pathogen and its interaction with the soybean plant and the seedbed environment. It is possible to view soybean planting failure in general as merely a collection of these many possible independent causes. However, a synthesis of information about individual causes is necessary both to understand the process of establishment in the field and to rationally apply practical control measures. In this paper, an attempt has been made to present a framework within which to interpret the overall epidemiology of soybean seedling diseases. To do this, soybean seedling establishment is considered first as a physiological process, then in terms of how seeds, pathogens and the environment interact in some individual diseases, and finally in terms of how manipulations of seeds, soil, and the seedbed microclimate can help lessen the probability of stand failure occurring in the field.

THE PROCESS OF SEEDLING ESTABLISHMENT

Water provides both the switch that activates the germination process and part of the medium in which germination takes place. A dry soybean seed at 11-12% water content has a water potential of well below -500 bars (349,440). This is too low to support the growth of even the most xerotolerant fungi (such as some *Aspergillus* spp.), which require water contents above 13% for growth (93,440). When a seed is planted, its water potential begins to equilibrate with that of the soil through imbibition. As water enters the seed, some previously solid materials in the seed enter into solution and metabolic processes become active (468). During initial imbibition, there is a relatively large outflux of materials from the seed into the soil (202,468). As the shift from an inactive to active seed is completed, the rate of exudation declines. A wide range of materials is released, including sugars, amino acids, inorganic ions and volatile organic compounds (187,202,203,468). Substrates and growth factors in seed exudate foster the growth of both pathogenic and nonpathogenic microorganisms in the seed and surrounding soil. If seed water potential reaches the range of -7 to -15 bars, cells in the radicle begin to expand and then divide (184,251). If the seed coat is intact, the protrusion of the radicle results in a peak in the rate of exudation and an increase in respiratory activity (47). As the seedling root elongates, the cotyledons are pulled upward away from the seed coat. The first part of the plant to break the soil surface is the crook area just below the cotyledon. Continued hypocotyl elongation elevates the cotyledons, and light induces the initiation of photosynthesis. As the cotyledons open, the first true leaves unfold and expand. The cotyledons may remain on the plant for a few days before falling to the soil surface. The complete transition from seed to established plant can take from about 8 days to over 3 weeks depending on soil temperature and moisture conditions.

Inoculum and the Environment

Although practically all soybean field soils contain inoculum of common damping-off fungi such as Pythium and Fusarium species, there is great variation in the incidence of most seedborne pathogens among soybean seed lots. For some pathogens, such as Diaporthe phaseolorum var. caulivora and soybean mosaic virus, seedborne inoculum is primarily of importance in initiating epidemics which spread among plants during the growing season. In such cases, even a low rate of seed transmission can be epidemiologically important (27,254). For other pathogens, such as Phomopsis longicolla, there is much more inoculum present in most production fields than can be introduced with seeds (177). Consequently, seedborne inoculum is primarily of importance only for the seeds which carry it.

Regardless of the amount of inoculum present, every pathogen has certain environmental requirements for growth. Additionally, most pathogens cause perceptible disease only under an even more restricted set of conditions. Work with seedborne and soilborne pathogens of a variety of crops has resulted in the general concept that the relative effects of different physical environments on a seedling disease can be understood in terms of the rates of host and pathogen growth under those conditions (184,303).

Pathogenesis by Phomopsis longicolla (sometimes referred to in the literature as Phomopsis sojae, Phomopsis sp., or Diaporthe phaseolorum var. sojae) is highly dependent on the physical environment. This fungus can be present in a large proportion of the seeds of some seed lots and has been of considerable concern in recent years (5,206,336,339,440,471,472,534). However, examination of the soil conditions under which P. longicolla causes damping-off has indicated that its economic importance may be more restricted than was initially thought.

For most infected seeds, P. longicolla is present only in the seed coat, although embryos can be killed or damaged if they are colonized before harvest (Gleason, M. L., and Ferriss, R. S., Unpublished data). Following harvest and storage, seeds which are infected, but still viable, can be killed by the pathogen either in the course of some seed quality tests or after planting in soil. In standard germination tests on rolled paper towels, seeds can be killed as a result of pathogen growth from seed coat or cotyleon infections, or infected seeds can be counted as abnormal because of the presence of mycelium. Such standard (normal) germination results are inversely correlated with the incidence of infected seeds among recently-produced seed lots (293,339). However, many seed lots which have a high incidence of infection and low standard germination can produce excellent stands in the field (163,206). This difference in performance is apparently related to the different physical conditions which a seed experiences on towels compared with in soil. Under soil conditions conducive to germination (as well as in sand-based germination tests), friction with the soil causes the seed coat to be left behind as the hypocotyl elongates. Because P. longicolla is present only in the seed coat in most infected seeds, this physical separation of the seedling and seed coat prevents subsequent colonization of living plant tissues by the pathogen. In contrast, there is relatively little friction between paper towels friction between paper towels and seeds. Consequently, the coat and seedling remain in contact for a longer period of time, and there is thus a greater chance that a superficially infected seed will be killed. Pathogenesis in soil appears to occur primarily in situations where seeds are planted into relatively dry soil at moderate temperatures (Fig. 1) (184). P. longicolla can grow well at water potentials lower than -100 bars (184). Consequently, if a seed which is infected only in its seed coat is planted in soil which is too dry to support germination, but wet enough to support growth of P. longicolla, it cannot escape the pathogen by germination until moisture conditions improve.

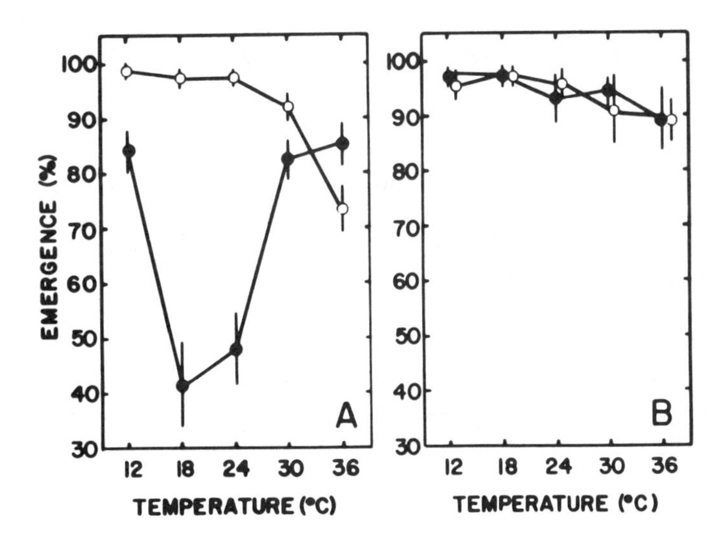

Fig. 1. Effects of initial soil moisture and temperature on emergence from soybean seed lots with low (-O-) and high (-●-) incidences of seed coat infection by Phomopsis longicolla in pasteurized soil. Seeds were incubated in soil for 3 days at -0.01 bars (A) or -15 bars (B) and the indicated temperature, and then for 18 days under near-optimum conditions for germination (M. L. Gleason and R. S. Ferriss, unpublished).

The occurrence of pathogenesis by P. longicolla only under certain planting conditions indicates that the inverse correlation of standard germination results with incidence of P. longicolla is in a sense an artifact of the testing procedure. The standard germination test is intended to predict seed lot performance under optimum planting conditions. However, interference by P. longicolla can make the test predictive mainly for dry, sub-optimum planting conditions.

Pathogenesis after planting by some other seedborne microorganisms also appears to occur mainly under certain soil conditions which do not support germination. Xerophilic fungi such as some species of Asperillus and Penicillium are primarily of importance during storage (93,440); however, these fungi can also be active on seeds in soil if the water potential is low enough (533). Similarly, Bacillus subtilis is associated with a wide variety of plant materials, including most soybean seed lots, but causes significant mortality only if imbibed seeds are incubated at a relatively high temperature (508).

SEED VIGOR

By considering the effects of the physical environment on pathogenesis in terms of host and pathogen growth rates, the general conclusion can be reached that the best overall conditions for planting soybeans are those which result in the fastest germination and emergence in the absence of pathogens. Considering the relative growth rates of host and pathogen can also help clarify the effects on establishment of seed quality characteristics other than the presence of seedborne pathogens.

Practically all soybean seed lots contain some seeds which have been mechanically damaged during production, harvesting, transportation, or storage. In some soybean seed lots, practically all of the seeds have been physiologically damaged from the physical weathering and microbial colonization associated with delayed harvest (562), or from the deterioration which takes place during storage (93,440). In standard germination tests, physiologically damaged seeds may germinate and produce normal seedlings; however, seeds from the same seed lot may fail to produce an adequate stand under some field conditions. This widely-recognized inconsistency between the performance of some seed lots under optimum and sub-optimum conditions has lead to the development of the concept of seed vigor.

The Association of Official Seed Analysts has adopted the following definition: "Seed vigor comprises those seed properties which determine the potential for rapid, uniform emergence, and development under a wide range of field conditions" (98). Although many other specific definitions have been proposed, there is a general consensus that good performance under sub-optimum planting conditions is a basic characteristic of high vigor seed lots (113,202,223,328). A number of seed quality tests have been developed which are intended to measure the relative performance of seed lots in stressful situations, and thus seed vigor (97,114,323,328,557). Both the cold test and the accelerated aging test have received a good deal of research attention. The cold test was originally developed to predict the suitability of corn seed lots for early spring planting (97). All variations of the cold test involve the incubation of seeds under wet, cold conditions, followed by incubation at near optimum conditions for germination. Most variations of the cold test involve the use of soil which is naturally infested with seedling pathogens such as _Pythium ultimum_. Results are expressed as the percentage of seeds which produced seedlings meeting standardized criteria. The accelerated aging test (also referred to as the artificial aging or controlled deterioration test) was originally developed to predict the relative suitability of seed lots for long-term storage, and involves the incubation of seeds at high temperature (usually 41 C) and high relative humidity prior to performing a standard germination test (114). Vigorous growth of some fungi and bacteria occurs during the temperature incubation period, and it is likely that the stress which is imposed is at least partly biological.

Among soybean seed lots, emergence and establishment in soil infested with _P. ultimum_ is correlated with results of both the accelerated aging and cold tests, but not with standard germination results (Fig. 2) (Ferriss, R. S., and Baker, J. M., Unpublished data). Conversely, emergence and establishment in pasteurized, uninfested soil are better correlated with results of the standard germination test than with the accelerated aging or cold tests. The different rankings of seed lots under conditions of high compared with low microbial stress confirm that vigor is an attribute of seed lots which is distinct from viability without stress. Although the exact nature of seed vigor is unknown, it is probable that it involves at least two interrelated phenomena which can affect microbial attack: seedling growth rate and the exudation of materials out of the seed (113,223,328,442,556,557). The more rapid germination and growth of higher vigor seeds may lessen the probability of death due to pathogenesis in the same manner that optimum planting conditions do: by resulting in a shorter period of exposure to seedborne and soilborne pathogens. The lesser amounts of nutrients which are usually released from higher vigor seeds provide a smaller amount of substrate for pathogen growth.

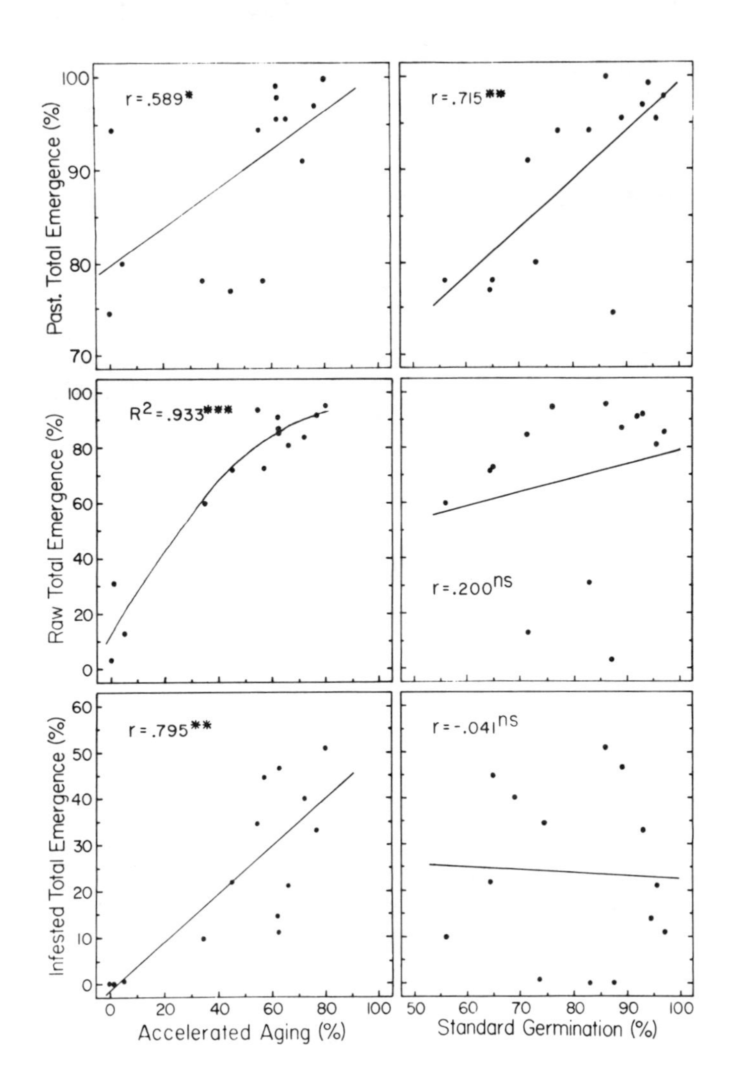

Fig. 2. Relationships between emergence of soybean seed lots in pasteurized soil, untreated soil and pasteurized soil infested with Pythium ultimum to results of the standard germination and accelerated aging tests (Ferriss, R. S. and Baker, J. M. unpublished). Significance levels for rho values: * = 0.05; ** = 0.01; *** = 0.001.

In addition to P. ultimum and high-temperature stress, higher vigor seeds and seedlings appear to have a greater ability to resist being damaged or killed by Phomopsis longicolla and Rhizoctonia solani (Ferriss, unpublished); however, the apparently biotic stress associated with soil flooding appears to be little affected by seed vigor. Preemergence soil flooding can have drastic effects on soybean establishment that are independent of those caused by

Phytophthora megasperma f. sp. *glycinea* (162,163). Seed and seedling mortality can be controlled to a limited extent by some seed treatments, and is probably due to the effects of low oxygen availability in combination with microbial activity during and after the flooding period. However, preemergence soil flooding appears to have similar effects on both high and low vigor seed lots. It is possible that the physiological process which is the basis of seed vigor requires aerobic conditions for its functioning, and, consequently, differences in vigor are not expressed during flooding.

In addition to its effects on stresses due to attack by microorganisms, seed vigor may affect the resistance of seeds and seedlings to abiotic stresses. Seedlings from higher vigor seeds may be able to break through soil crusts more easily, and the more rapid early growth of their roots may allow them to better withstand the effects of post-germination drought.

CONTROL

Seedling establishment is but one small part of any system of soybean production, and situations where low stand significantly affects yield appear to be relatively rare and unpredictable. Consequently, economically realistic control measures for establishment failure must be inexpensive and/or provide benefits in addition to increased establishment. Considering what control measures to use is analogous to considering the purchase of an insurance policy: the possible events for which it is most important to obtain coverage are those which have a relatively high probability of occurrence and/or can have a devastating effect if they occur. As with any disease, the grower must make a judgment as to whether the costs of controlling establishment failure are justified by the expected benefits. Unfortunately, there are no actuarial tables available with which to estimate the likelihood of significant yield loss due to the lack of control.

The most obvious and effective control measures for seedling diseases involve the seeds which are planted. Although there is much evidence that seed lots with high vigor and a low incidence of seedborne pathogens are much less likely to experience problems with stand establishment, results of vigor tests and pathogen assays are rarely available to growers. Until this situation changes through market forces or regulation, a soybean grower must rely on other information to select high quality seed. In particular, it can be recommended that seed lots be planted which are known to have high standard germination results, to have a low incidence of moldy seed, to have been produced in the previous growing season, and to have been stored under reasonable conditions of temperature and humidity. In addition, some wide-spectrum chemical seed treatments can provide protection against many of the same diseases which are affected by seed vigor, as well as some other ones such as damping-off due to preemergence flooding. In cases where low quality seeds must be used, seed treatment can provide an easily justified measure of insurance (23,206,532). However, the benefits of treating high quality seeds are less certain, except where treatment is an integral part of a Phytophthora root rot control strategy (448). The likelihood of a significant amount of establishment failure occurring is much less for high quality seeds than for low quality seeds. Consequently, treatment of high quality seeds may not be economically justified. Until more is known about the likelihood of occurrence of particular adverse environmental conditions (such as soil flooding) and the possible detrimental effects of seed treatments (such as slowed emergence and interference with beneficial microorganisms), there is little rational justification for most uses of seed treatment on high quality seed.

The soil into which soybean seeds are planted provides both a source of pathogens and the environment in which early germination and seedling growth take place. Although the inoculum density of soilborne seedling pathogens can be affected by crop rotation, the wide host ranges of many of them (such as *Pythium*, *Rhizoctonia* and *Fusarium* species) make it difficult to judge the effects of different rotation crops. Consequently, most management practices which are intended to affect establishment operate through effects on the microclimate encountered by planted seeds. Under dry conditions, relatively deep planting can place seeds in contact with moist soil; however, deep planting also delays emergence and thus results in a longer period of susceptibility to pathogens. In many areas, delaying planting until after cold, wet conditions are likely to occur can result in faster emergence and less establishment failure. Seedbed preparation practices such as soil ridging can also increase the temperature experienced by seedlings and thus increase the probability of establishment.

Although a number of control measures can be used to control soybean seedling diseases, the economic benefits of maximizing establishment are not clearly known. Soybeans have a great ability to compensate for the early-season death of adjacent plants (80), and little or no significant yield increase is usually observed in response to seed treatment and the use of high quality seed (23,206,532). However, the effect of plant population on yield can vary in different production environments (105,106,261). Control of soybean seedling diseases may have its greatest economic impact in allowing predictable stands to be obtained, rather than in maximizing stand density. If almost all planted seeds can be expected to produce established plants, then it is easier to obtain intermediate stand densities which are high enough for optimum yield, but not so high as to increase the risk of lodging (105,261). Future improvements in estimates of the benefits of controlling soybean seedling diseases depend in part on the development of a better understanding of the effects of plant population on yield.

SEED TREATMENTS AND THE FUNGAL PATHOGENS THEY ARE DESIGNED TO CONTROL

Ademir Assis Henning

EMBRAPA/CNPSoja
Londrina, PR - Brazil

Soybean, Glycine max (L.) Merrill, has become one of the major sources of oil and protein. In 1980, the USA produced 66% of the total world soybean production, followed by Brazil (16%) and China (9%) (470). A considerable increase in soybean production was observed during the last three decades because of a rapid expansion in the traditional growing areas and the introduction of this crop into new areas (334). With this rapid expansion, diseases also increased in number and severity. Significant increases occurred in tropical and subtropical zones, where seed quality is a major problem (171,208).

Mechanical damage, adverse weather conditions, insect damage, improper storage, and seed-borne fungi, can all reduce seed quality (171,207, 210,339,459,532). Fungicide seed treatment is used commercially to some extent in the USA and Brazil to improve germination of infected seeds, to reduce the amount of seed-borne inoculum, and most importantly, to protect seed and seedlings from soil-borne pathogens (334).

Soybeans in the field can be attacked by a number of fungi, bacteria, viruses and nematodes. Because of their great number and the losses they induce in yield and seed quality, fungi are considered to play a major role in limiting soybean production (171). According to their characteristics, these organisms can be classified as seed-borne, soil-borne, and storage fungi.

SEED-BORNE PATHOGENS

Phomopsis sp. [Diaporthe phaseolorum (Cke. & Ell.) var. sojae Leh.] Pod and Stem Blight

This pathogen, among all those that invade seed in the production field, is considered to be the most prevalent fungal organism in seeds (177,208,364,394,532). A humid and warm environment during seed maturation, as well as harvest delay, can markedly increase seed infection (171,208,364,459,470). Severely infected seeds are shriveled, have a white, chalky appearance, and usually do not germinate (334). In temperate zones, under conditions that favor Phomopsis seed infection, the fungus can be controlled by benzimidazole fungicides applied to the seed crop at growth stage R6 (full seed). Predictive systems for foliar fungicide application have been developed (335,483). Depending on environmental conditions during the growing season, seeds with normal appearance may have high percentages of Phomopsis sp. (208). In seed of this type, the fungus can drastically reduce germination in laboratory tests (warm germination), but may not affect field emergence if physiological quality of the seed is good. One possible explanation for this may be that infection, occurring late in the season due to climatic conditions favorable for the pathogen, results in internal but not deep seated seed-coat infection. As observed in 1980 in Brazil (208), seedlots that had reduced germination in laboratory tests due to Phomopsis sp. had high vigor and viability as determined by the tetrazolium (2,3,5 triphenyl tetrazolium chloride) test. These results correlated positively with field emergence of nontreated seed. Seed treatment with thiabendazole or captan controlled Phomopsis sp. and improved germination in the laboratory (208). It was also observed that germination was not increased by fungicide seed treatment in seedlots with reduced physiological quality caused by weathering, insect damage, or mechanical damage. Similar results have been reported by other authors (459,532).

Diaporthe phaseolorum (Cke. & Ell.) var. *caulivora* (Athow and Caldwell)
Stem Canker

This pathogen can be seed- and crop residue-borne and also is capable of killing plants between mid-season and maturity (334). A recent outbreak of the disease in the southern USA caused severe losses in 1981. Proper identification of the pathogen on infected seed can be made only by seed health tests because the symptoms on seeds are quite similar to those of Phomopsis sp. (caution must be exercised because infected seed can also be symptomless). Seed treatment with fungicides may be of value, but the role of infected seeds as a source of inoculum has not yet been determined (334). However, it is important to prevent the long-range dissemination and introduction of this potentially dangerous pathogen into disease-free areas.

Colletotrichum truncatum (Schw.) Andrus and Moore
Anthracnose

Symptoms are commonly seen towards the end of the season as irregular black blotches on senescent stems (334,451). However, in tropical and subtropical regions, this disease is serious during the growing season. Frequently petioles, and sometimes stems, are killed prematurely by the fungus (170,171). Seed infection, which is commonly low in temperate zones (334), may be very high in the tropics (366) where levels above 50% have been reported by Franca Neto et al. in Brazil (170) and Agarwal (11) in India. Seed treatment with fungicides may be of value to reduce seed-borne inoculum and limit the spread of this pathogen into new areas in the tropics.

Cercospora kikuchii (Mats. & Tomoy.) Gardner
Purple Seed Stain

Although the typical symptom of seed infection is the purple discoloration of the seed coat, not all infected seeds become discolored. Some reduction in seed germination has been reported (456,548), but Franca Neto et al. (169) observed no reduction in seed quality (germination and field emergence) or yield of three soybean cultivars with levels of up to 40% purple-stained seeds (Table 1). Seed-borne inoculum levels are not often correlated with harvest losses (169,339,548). Seed treatment may have some value as a preventive measure to reduce the source of initial inoculum. The fungus overwinters in crop residues and a foliar blight phase may affect yield (535).

Table 1. Effect of five levels of purple-stained seed on seed quality and yield of three soybean cultivars. EMBRAPA-CNPsoja. Londrina PR, Brazil.

Cultivar	% PSS[1]	% Viable *C kikuchii*[2]	Warm Germ %[3]	% Emergence Under Natural Field Conditions	Yield kg/ha	*C kikuchii*[4] %
Parana	0	0.12	84.75abc[5] B	81.19[6] ab A	1730 NS[7]	1.25
	5	0.25	87.37a A	83.69 a A	1520	0.87
	10	1.87	86.38ab A	80.94 ab A	1995	0.50
	20	5.62	82.50 c B	83.12 ab A	1886	1.62
	40	11.37	83.16 bc B	77.81 b B	1676	0.50
Davis	0	0	79.12a C	79.50 a A	2020	1.00
	5	2.37	78.50a B	75.81 ab B	2043	1.87
	10	5.87	77.58a B	75.94 ab A	1762	0.75
	20	14.62	81.37a B	75.94 ab B	1940	1.62
	40	29.37	80.87a B	73.38 b B	1655	0.50
Bossier	0	0.25	93.12a A	81.62 a A	2079	1.87
	5	3.50	88.75 b A	76.75 ab B	1897	1.00
	10	6.25	88.12 b A	81.25 ab A	2240	2.12
	20	15.25	91.75ab A	79.62 ab AB	2181	1.25
	40	25.62	89.75ab B	84.19 a A	1972	1.62
C.V. %			2.87	4.36	17.52	

[1]Percentage of purple-stained seed (visual rating).

[2]Percentage of viable *C kikuchii* as determined by the blotter test (four replications of 200 seed per treatment).

[3]Percentage germination in rolled paper towel at 25 C for 7 days.

[4]Percentage of infected seed after harvesting as determined by blotter test.

[5]Means separated by the Duncan's multiple range test at the 5% level of probability (small letters compare purple stain levels inside cultivar and capital letters compare cultivar inside each level of purple-stained seeds).

[6]Data transformed in arc-sin $\sqrt{\%}$.

[7]No significant differences ($P \geq 0.05$).

Cercospora sojina Hara
Frog Eye Leaf Spot

Infected seeds may show a greenish-grey discoloration (171) and may affect germination (456). Although infected seed is not considered to be a major inoculum source (334), seed treatment with thiabendazole at (0.2g) or thiram (1.4 to 2.1g) of active ingredient per kg of seed is currently recommended in Brazil (210). This is considered an important preventive control measure (besides cultivar resistance), since this disease can be severe on susceptible cultivars in the "*cerrado*" area.

MISCELLANEOUS

Peronospora manschurica (Naum.) Syd. causes downy mildew. Infected seeds, encrusted with oospores of the fungus, may be an important initial source of inoculum (363). Seed treatment may give some control but resistant varieties are the most effective control measure (334) to control secondary conidial spread on foliage. In addition to *Septoria glycines* (Hemmi), the brown spot organism, other foliar pathogens less frequently found in seeds include *Corynespora cassiicola* (Berk. & Curt.) Wei. (target spot) and *Myrothecium roridum* (Tode ex. Sacc.). Their effects on seed quality are negligible and infected seed may be important only as a source of initial inoculum, where seed treatment, as a preventive control measure, could have its place (171).

SEED- AND SOIL-BORNE PATHOGENS

Sclerotinia sclerotiorum (Lib.) deBary
White Mold

This fungus can cause serious disease in some of the best seed-producing areas of southern Brazil (171). The fungus can be transmitted both as sclerotia mixed with the seed and as internal, seed-borne dormant mycelium (363). The latter, although less common, can be responsible for the long range dissemination and introduction of this organism into new areas. Once introduced, the organism is difficult to eradicate, since its resting structures (sclerotia) can remain viable in the soil for many years (171,363). For this reason, seed treatment with thiabendazole or thiram is recommended for seed produced in fields with trace amounts of the disease. Heavily contaminated fields should not be used for seed production (210).

Macrophomina phaseolina (Tassi) Goid.
Charcoal Rot

The disease caused by this fungus is common in soybean fields growing under moisture stress. The fungus can kill plants prematurely and thus reduce yield (seed size) and seed quality (171). Infected seeds normally do not germinate (171) and field emergence may be reduced (174). However, the low rate of seed transmission and its presence in most soils reduces the importance of its seed-borne phase. Seed treatment may control the fungus but will not protect plants late in the season.

Fusarium spp.

Fusarium spp. are frequently associated with reduced germination of soybean seeds (208,364). In Brazil, *Fusarium semitectum* Berk is the most prevalent species in seeds. Seed treatment with captan or thiabendazole resulted in higher germination (207,208). *Fusarium oxysporum* Schelct, and *F. solani* Mart. are also common damping off organisms in Brazil.

Rhizoctonia solani Kuehn

Rhizoctonia solani Kuehn is also a common damping off pathogen in Brazil.

Seed treatment with a number of fungicides gives good control of *Fusarium* and *Rhizoctonia* spp. and particularly improves field emergence in soils with low moisture content (dry soils).

SOIL-BORNE FUNGI

Phytophthora megasperma f. sp. *glycinea* Kuan and Erwin, one the most destructive root rot diseases of soybean in the USA, can kill plants throughout the growing season (334). One problem with recently introduced varieties tolerant to *P. megasperma* is their susceptibility to seedling damping-off by *Phytophthora*. This problem can be reduced however, by seed treatment with metalaxyl (394), which can give good control in the first few weeks after planting (334).

Other important soil-borne pathogens such as *Pythium debaryanum* Hass and *Pythium ultimum* Trow are also controlled with metalaxyl, PCNB (334), or pyroxyfur (394).

STORAGE FUNGI

Several species of *Penicillium* and *Aspergillus* can invade soybean seeds that are stored at moisture contents above 14%. Storage losses are aggravated in the tropics where temperatures above 28 C and relative humidities of 85% or more are common. Besides physiological aging, these conditions favor the development of *Aspergillus flavus* which cause complete seed deterioration (171). Seed treatment with fungicides may control the fungus; however, it is of little value because of the physiological aging which occurs under such extremely adverse conditions.

EFFECT OF SEED TREATMENT ON FIELD EMERGENCE AND YIELD

It is well-documented that seed treatment with fungicides may increase emergence and plant stands. Yield increases, however, are less frequently reported (23,72,90,208,364,394,459,532,553). This lack of yield response in experimental plots is often attributed to the compensatory growth of soybeans and reduced weed competition (344). In large fields, soil preparation (seedbed) and weed control may not be as adequate as in experimental plots. These factors may result in larger differences between treated and nontreated seeds than one normally finds in experimental plots. Furthermore, when soybean fields with very low plant populations are combine-harvested, weeds and lower pod height may increase harvest losses (Table 2). The effects of plant population on harvest losses have been demonstrated further when the cultivar 'Bossier' was planted in very dry soil in Londrina, Brazil. Yields from high population plots (treated with thiabendazole) versus low population (control) were 28.8% or 49.7% greater, respectively, from hand-harvested than from machine-harvested plots. These results demonstrated the importance of seed treatment when sowing is done in dry soil in tropical and subtropical areas.

Table 2. Effect of fungicide seed treatment on population, plant height and yield of 'Bossier' soybean planted on dry soil in Londrina-PR, Brazil. EMBRAPA-CNPsoja.

Treatment	Plant population (plants meter^{-1})	Height	Yield (12% moisture) Manual[5] kg/ha	%	Combine[6] kg/ha	%	Harvest losses %
Thiabendazole	17.6[2] a[3]	49.2[4] a	1347 a	41	1046	64	22.4
TCMTB[1] (Busan 30E)	7.8 b	38.6 b	995 b	4	698	9	29.8
Control	8.8 b	41.0 b	955 b	0	638	0	33.2
C.V. %	21.16	8.2	20.6				

[1] 1,2-(thiocyanomethylthio) benzothiazole.

[2] Average of five replications (four center rows, five meters long) randomly assigned to each treatment.

[3] Means separated by the Duncan's multiple range test at 5% level of probability.

[4] Average of 10 plants measured at random in each plot, with five replications.

[5] Yield based on the 4 center rows of each plot, with five replications (hand harvested and threshed).

[6] Yield of the combine-harvested and threshed area of 1219 m^2 per treatment. Results not analyzed because of no replication.

Gilioli et al. (182) reported that increasing planting depth as the soil moisture decreases can result in better emergence, at least in some types of soil. This practice was further evaluated at three different locations in Brazil; the results (Table 3) show the risks involved in deeper plantings. The best practice is to plant at the conventional depth (4-5 cm) and use fungicide treatment when planting in dry soil.

Table 3. Effect of planting depth and seed treatment on seedling emergence of soybean 'Parana', planted at three different localities under distinct edapho-climatic conditions. EMBRAPA-CNPSoja.

Fungicide Treatment[1] and Planting Depth	Seedling emergence[2]		
	Londrina[3]	Ponta Grossa[4]	Sao Miguel do Iguacu[5]
4 cm no fungicide	14.5 c[6]	73.9a	86.3a
4 cm with fungicide	88.5a	68.1 b	85.8ab
8 cm no fungicide	58.3 b	48.2 d	81.2 c
8 cm with fungicide	84.5a	57.4 c	82.7 bc
C.V. %	5.4	7.6	3.1

[1]Thiabendazole, a.i. 0.2 g/kg of seed.

[2]Data transformed in arc-sin ($\sqrt{\%}$), final counting on the 28th day.

[3]Dry soil.

[4]Hard surface crusting formed after heavy rain.

[5]Soil at the field capacity.

[6]Means separated by the Duncan's multiple range test at the 5% level of probability.

Although soybean seed treatment is not widely used (334), the conditions under which it may be beneficial are quite well established in the USA (23,334,364,459,532). Basically, seed treatment is recommended when planting poor quality seed; when adverse conditions, such as cold wet soil, occur in the seedbed at planting time; when reduced seeding rate is used; or when the field is planted for seed production (459). In Iowa, a detailed study conducted between 1980 and 1983 (334,532) provided the basis for soybean seed treatment in that state, based on seed quality, planting date and soil temperature (334).

In Brazil, the recommendations for seed treatment are similar to those discussed above in many respects. Major differences arise when attempts are made to apply results from the United States to tropical and subtropical regions. Current recommendations are based on a large number of field experiments conducted between 1979 and 1985 in the major soybean producing areas under different edaphoclimatic conditions. Basically, seed treatment is recommended when planting occurs in dry soil (a common situation in central Brazil); seed of medium to low vigor has to be used; and when planting is done in cold wet soil (particularly in the low rice lands of the south). Under any of these conditions, seed germination and seedling emergence rates may be reduced. Consequently, seed remaining longer in the soil are exposed to soil-borne pathogens such as Rhizoctonia solani, Fusarium spp. (mainly F. semitectum and Aspergillus flavus that can cause serious stand establishment problems, particularly in tropical and subtropical areas (171). Thiram, thiabendazole, and captan are the most widely used fungicides although carboxin, carboxin + thiram, PCNB + captafol and methyl thiophanate + thiram are also labelled and used to some extent in Brazil (personal communication).

In conclusion, seed treatment of soybeans is used to a limited extent in the USA and Brazil for controlling seed-borne Diaporthe phaseolorum var. caulivora, D. phaseolorum var. sojae, Phomopsis longicolla Hobbs sp. nov. (in the USA). Colletotrichum truncatum, Cercospora kikuchii C. sojina Fusarium spp., and Sclerotinia sclerotiorum (in Brazil). Soil-borne fungi controlled by seed treatment in the USA are Pythium debaryanum, P. ultimum and Phytophthora megasperma var. glycinea. Other common damping off organisms in Brazil, like Fusarium oxysporum, F. solani, and Rhizoctonia solani are controlled by several fungicides.

EVALUATION OF CURRENT PREDICTIVE METHODS FOR CONTROL OF PHOMOPSIS SEED DECAY OF SOYBEANS

D. C. McGee

Department of Plant Pathology, Seed Science Center
Iowa State University
Ames, IA 50011

Phomopsis seed decay caused by *Phomopsis longicolla* and *Diaporthe phaseolorum* var. *caulivora* is a major concern to soybean seed growers because of adverse effects on seed germination. Control can be achieved by applying funcigides to the growing crop. Very often, however, the disease is not severe enough to justify application costs. This paper discusses two predictive methods that identify fields that might benefit from spraying. Both methods are used in conjunction with fungicides labelled for application at R6 (151); these are all in the benzimidazole group and include Benlate 50W, Mertect 340F, and Topsin M 70W.

DEVELOPMENT OF PREDICTIVE METHODS

In the mid-1970s, research on the application of benzimidazole fungicides to soybeans indicated good control of seed infection by *Phomopsis* and *Diaporthe* spp. and improvement in yield (142,402). The fungicides Benlate 50W (benomyl) and Mertect 340F (thiabendazole) were introduced as foliar sprays to be applied to soybean crops at the R3 growth stage and 14 days later. Labels claimed control of several diseases, including pod and stem blight, purple seed stain, anthracnose, and Septoria brown spot. The primary incentive was to increase yield. No distinction was made for use on seed crops.

It soon became evident, in northern production areas, that yield responses were inconsistent. A "point system" method, therefore, was developed in Illinois (465) as an aid in identifying fields that would benefit from spraying. Other states followed with modifications of the system. These models were essentially intuitive and lacked a sound epidemiological base. They proved ineffective, and, in northern areas, the use of foliar fungicides to increase yield essentially has been abandoned.

Attention was then focused on the use of foliar fungicides on seed crops for control of Phomopsis seed decay. One application of a fungicide, at growth stage R6, was shown to be effective (506). Disease severity varied greatly from year to year, and fungicides were not always needed. It was clear that predictive methods would be of value. Researchers in Kentucky developed a "point system" method specifically designed for seed crops (482,507). Another method was developed in Iowa using pod infection by *Phomopsis* and *Diaporthe* spp. as an indicator of subsequent seed infection (335,340).

EPIDEMIOLOGICAL BASIS OF PREDICTIVE METHODS

The Illinois predictive system utilized ten criteria considered to affect either the diseases or crop value. Points were assigned to each criterion, and higher totals indicated greater risk of crop losses (465). The Kentucky system is based on four of these criteria: cropping history, varietal selection, planting date, and rainfall. All these have been shown to be associated with increased severity of Phomopsis seed decay. Garzonio and McGee (177) showed that seed infection was greater in fields in continuous soybean than in those where a corn-soybean rotation was used. Early-maturing varieties or early-planted crops are at higher risk of developing Phomopsis seed decay because seed maturation occurs when warm, wet weather, favorable for seed infection, is most likely to occur (34,505). Two studies (297,505) indicated that rainfall between flowering (R2) and physiological maturity (R7) favors disease development (297,505). In 1984, refinements were made within the various criteria, and the predictive model was developed for use in a microcomputer (D. Hershman, University of Kentucky, pers. commun.).

The Iowa predictive method is based on the fact that pods are a pathway for seed infection (281). Pod infection is measured by detaching pods at the R6 growth stage in the field, treating with a surface sterilant (sodium hypochlorite), immersing them in the herbicide, Basagran, and incubating in a moisture chamber for seven days. Characteristic fruiting bodies of the pathogen develop on the pod surface, and the number of infected pods is used as the predictive measurement (340). For the method to be feasible, it had to be shown that seed infection could not occur before R7 (335); this was important because fungicides are ineffective after seeds are infected (338). It was also important to determine whether inoculum reaching pods after pod infection was measured (i.e., R6) could cause seed infection. Field and growth-chamber experiments indicated that this was unlikely to occur in northern states such as Iowa (335).

The Kentucky method was tested on 19 fields in 1982. It accurately identified six fields that did not require spraying and five that did. The remaining eight fields had intermediate point totals, and seed infection levels were not accurately predicted (507). McGee (335) showed a correlation coefficient of 0.68 between pod infection values and subsequent seed infection in 23 soybean fields in Iowa in 1982. Even better validation of pod infection as a predictive measurement, however, was reported from Kentucky (507) where it was well correlated (r=0.73) with seed infection in 38 fields over a four-year period.

LIMITATIONS OF PREDICTIVE METHODS

Although the Iowa and Kentucky methods were developed on a sound epidemiological basis and have proved practical, they do have limitations. The Kentucky method relies entirely on indirect estimates of disease severity and cannot account for unusual circumstances that may favor disease development. Furthermore, although there is no question that increased Phomopsis seed decay is associated with each of the criteria used in the method, there is no experimental data to show their combined effects on disease severity. The point values given to each criterion also are empirical. A major advantage of the method is that it is based on information available at the time fungicides are to be applied. Also, no sampling or laboratory work is required.

The pod test method measures the inoculum that has the potential to move to seeds. It, therefore, is a more reliable means of detecting unusual increases in disease severity. Equipment and materials are inexpensive and readily available (340), and seed company personnel are able to carry out the test with minimal training. Disadvantages are that pods have to be sampled and processed. Also, laboratory or office space is needed to incubate and read the tests. The incubation period is seven days leaving only seven more days to take any action to apply the fungicides. There are obvious logistical problems in testing large numbers of fields in such a short period. Several Iowa seed companies, however, have demonstrated that, with good management, this can be accomplished. Another option, often used, is to test only "high risk" fields that might indicate the overall level of infection in the region. The criteria in the Kentucky method are used to detect these fields (340). A problem with this approach is that it is based on the assumption that all varieties are equally susceptible to Phomopsis seed decay.

Tolerance levels exist within both methods that indicate whether or not to apply fungicides. There are also levels that are indefinite. For fields in these categories, the decision to spray seems to be best resolved by using features of both methods. Tekrony et al. (507), for example, improved the overall predictability of the Kentucky method by incorporating pod infection data for fields with marginal point total. McGee and Nyvall (340) suggested that date of maturity and short-term weather forecasts be used in determining whether fields with pod infection in the 25-50% range should be sprayed.

A definite weakness in both methods is that they cannot account for the effect of weather after the predictive measurement has been made on seed infection. As discussed, the pathogen will not infect seeds until the R7 growth stage (281,335). Seed infection may then occur when there is precipitation that will maintain the relative humidity close to 100% for prolonged periods between R7 and harvest. Balducchi and McGee (34) showed that the higher the temperature and the younger the plants during such wet periods, the shorter is the time needed for seed infection. Rupe and Ferriss (430), suggested that the plant age effect is related to pod moisture. Findings in these papers provide a fairly precise definition of climatic conditions that favor seed infection.

USE AND EFFECTIVENESS OF PREDICTIVE METHODS

The Kentucky point system was published in 1981. Current estimates indicate that 20% of soybean seed growers in Kentucky consider fungicide spraying, and 20% of these use the predictive method (D. Hershman, pers. commun., 1986). In Iowa, cooperative experiments were carried out with seed companies during the development of the pod-test method. This gave industry personnel direct experience with it. It is now used by several companies in different states. They use it on "high risk" indicator fields, on fields with valuable product lines, or on all production fields. Informal surveys of companies that previously sprayed with a fungicide on a regular basis indicate savings in spraying costs of more than $1,000,000 in Iowa since 1983. This was accomplished without any obvious losses in seed quality.

FUTURE USE OF PREDICTIVE METHODS

The two predictive methods described in this paper have been in use for several years and have proved effective. Factors influencing any plant disease are complex, however, and predictive methods often are reliable only within a particular range of environmental conditions. McGee (335) demonstrated that, if pods were exposed to prolonged periods of wet weather accompanied by unusually high temperatures, inoculum of *Phomopsis* and *Diaporthe* spp. reaching pods after the predictive measurement had been made could cause seed infection. These conditions occurred in Iowa in 1986, and the pod test did not predict significant seed infection in some fields (McGee, unpublished). Although these instances can be explained by extreme weather, they do, however, point up the importance of improving the method by further research. Work is now in progress in Iowa to design a mathematical model that will incorporate pod-infection values and weather forecasts. Efforts are also under way to develop a test to reduce the time needed for the pod test from 7 days to a matter of hours by using monoclonal antibody technology. There already is good evidence that each predictive method can be improved by incorporation of features from the other. Further work on combining them could prove beneficial.

MOLECULAR APPROACHES TO THE DETECTION OF SEED-BORNE PLANT PATHOGENS

John H. Hill

Depts. of Plant Pathology, Seed and
Weed Sciences, and Microbiology
Iowa State University
Ames, IA 50011

Traditional approaches to detection and quantitation of seed-borne pathogens have included grow-out tests, plating seeds on microbiological media, incubation tests in moist chambers or on blotters, and various serological tests. Some tests have been highly successful, whereas others are of dubious accuracy and value. Many can be labor intensive, and, because of the time required to perform a test, results are often not obtained in timely fashion. Recently, application of some tools of molecular biology has proved useful for detection of pathogens in seed. Many of the most recent advances have been made with detection of viruses. Techniques for detection of other pathogens, using principles similar to those developed with viruses, are rapidly being developed. Because many of the initial approaches have been developed using virus or virus-like agents as models, we will consider primarily the detection of viruses or virus-like agents in seed.

A number of viruses which cause significant economic loss are seed borne. Infected plants derived from this seed can provide a randomly distributed primary inoculum source for subsequent secondary spread of such viruses. Furthermore, seed trade provides a ready means for spread and international distribution of seed borne viruses into regions where they had not previously existed. This is particularly important with the current emphasis on exchange and importation of germplasm. Several molecular approaches are being used or developed for virus detection in seed. These include application of polyclonal and monoclonal antibodies in immunosorbent assays, complementary DNA (cDNA) technology, and electrophoresis. This discussion will focus on these applications of molecular biology to detection of seed-borne pathogens.

Serological assays have been used for detection of plant viruses for many years. The utility and sensitivity of these assays received significant impetus with the development of immunosorbent assays such as enzyme-linked immunosorbent assay (ELISA) and radioimmunosorbent assay (RIA). Numerous examples can be cited where both ELISA and RIA have been used for detection of seed-borne viruses (65,66,116,181,233,234,311). Traditionally, most of these have utilized polyclonal antisera produced in rabbits. These antisera contain polyclonal antibodies or an array of antibodies specific to different antigenic sites (epitopes) on the coat protein of a virus. Antibodies are proteins, found in the blood serum, which belong to the group of immunoglobulins capable of binding specifically to the antigen. Such serum can be difficult to produce for viruses that are laborious to obtain in purified form and frequently can react with "healthy plant tissues" to confound, invalidate, or reduce sensitivity of assays. Cross absorption and extensive antibody purification can be used to obviate these difficulties (341). However, because of the polyspecific nature of polyclonal antiserum, it is virtually impossible to insure that different preparations of antiserum to the same virus have identical specificity. Therefore, once a particular antiserum preparation with known specificity is exhausted, it is improbable that the preparation can be precisely duplicated.

Immunological research was dramatically redirected when Kohler and Milstein (284) showed that somatic cell hybridization could be used to develop a continuous cell line capable of producing monoclonal antibody. Monoclonal antibodies are produced by specialized cell hybrids called hybridomas. Hybridomas are made in the laboratory by fusing myeloma tumor cells, which can efficiently replicate in laboratory culture, with antibody-producing cells from the spleen (spleenocytes, which are not amenable to endless cell culture). The hybridomas produce antibodies and grow continuously in culture. After appropriate isolation of individual colonies from individual cells, desired cell lines are grown in quantity. Initial selection of appropriate hybridoma clones is best done by using assay methods similar or identical to those for which the monoclonal antibody will be used. Clones producing desired antibodies can be frozen in liquid nitrogen ensuring a source of antibodies for the future, which are identical to that originally obtained. Additionally, in contrast to generation of polyclonal antibodies, very small amounts of antigen (ca. 50 μg) are needed to produce monoclonal antibodies.

Monoclonal antibodies can be produced from hybridomas grown in culture, or alternatively, hybridomas are injected into the peritoneal cavity of a mouse, causing a tumor. The tumor cells secrete large quantities of antibody into a serum-like fluid called ascites fluid. In contrast to the polyspecific nature of polyclonal antibodies, monoclonal antibodies are specific to a single epitope. Monoclonal antibodies selected to detect all strains of a particular virus should react with strain-specific epitopes. Therefore, antibody selection and screening methods are very important for selection of relevant antibodies.

Polyclonal and monoclonal antibodies have both been used as diagnostic reagents in immunosorbent assays. The use of polyclonal antibodies, although well established for detection of plant viruses, can have limitations as discussed previously. Monoclonal antibodies also can have limitations. Unexpected cross-reactivity with proteins considered to be unrelated to viral proteins such as cytochrome C or ovalbumin can occur (C.-A. Kubanek and J. H. Hill, unpublished results). If antibodies are selected for a strain-specific rather than a virus-specific epitope, detection of all strains of a virus will not be possible. Additionally, monoclonal antibodies can also be cross-reactive with several different viruses, generally within the same virus family. Obviously, this can preclude accurate detection of a single virus. Fortunately, unique characteristics of individual monoclonal antibodies generally reduce the importance of such potential limitations and emphasize the necessity for careful screening directed toward the use for which the antibody is desired. The unique character of each monoclonal antibody, in fact, provides the opportunity for exquisite manipulation of assay systems that is not possible with polyclonal antibodies.

Although many variations of immunosorbent assays exist (37,283), most investigators in plant pathology have favored the "double sandwich" form of ELISA or RIA (65,91). In this assay, a "capture antibody" employed to coat a solid phase generally made of polystyrene plastic (or other materials, such as the nitrocellulose employed in "dot-blot" assays (36)), is used to trap the virus. Another antibody, generally conjugated with an enzyme, is used as a second antibody. Previous studies (117,282) have suggested that direct conjugation of the second antibody with an enzyme can induce molecular changes that may alter the avidity and/or affinity of the antibody molecule for the virus and induce a high degree of strain specificity, which may be undesirable in some circumstances. These problems can be circumvented when monoclonal antibodies are employed, as we shall shortly see. Alternatively, if polyclonal antibodies are employed in the assay, it is necessary to have available specific antisera prepared in different animal species or prepare $F(ab^1)_2$ fragments and use Fc-specific reagents (37,283). These manipulations are often inconvenient or impractical.

Because most workers are now familiar with the use of polyclonal antisera in immunosorbent assays, the remainder of this discussion on these assays shall be confined to the use of monoclonal antibodies. Early research in our laboratory suggested that the most sensitive double-sandwich immunosorbent assays require the use of epitopically distinct monoclonal antibodies. This has been demonstrated in experiments in which the second antibody was labeled with either tritium (230) or biotin (H. I. Benner and J. H. Hill, unpublished). In a similar fashion, a mixed polyclonal-monoclonal antibody system may be unsuccessful if the polyclonal antibody is used as a capture antibody and the monoclonal antibody is used as a second antibody (230). The converse is not true, however. The existence of epitopically different monoclonal antibodies can be demonstrated by competitive assays (117) or by probing Western blots of polypeptides derived from partial proteolyzed viral coat protein with different monoclonal antibodies (263).

To avoid potentially undesirable effects caused by direct enzymatic conjugation of the second antibody, we have examined a procedure using a second monoclonal antibody labeled with biotin in a double-sandwich ELISA (116). The small size of biotin (mol. wt. = 244) and the gentle conditions required for coupling it to proteins (40) make biotin an attractive marker for immunoglobulin molecules. Biotinylation allows binding of several biotin molecules to a single protein (246), and, even after extensive substitution of amino groups in antibody molecules by biotin, antigen-binding capacity is not modified (116,198). The detection system is then dependent upon the strong noncovalent binding between enzyme-labeled avidin or streptavidin and biotin ($Kd=10^{-15}M^{-1}$). The strength of this interaction can be assessed when compared with the antigen-antibody interaction ($Kd=10^{-5}-10^{-9}M^{-1}$).

An alternative approach, and perhaps the most desirable with respect to absence of potential antibody modification by conjugation, once again utilizes the unique potential of well-characterized monoclonal antibodies. Because antibodies are immunoglobulins, which are divided into five different classes (525), it is possible to use the molecular differences among immunoglobulin classes to advantage. Immunoglobulin G (IgG) has been most useful in studies with plant viruses, whereas immunoglobulin M (IgM) has played a secondary role. Recently, an ELISA to specifically differentiate maize dwarf mosaic virus strains A and B on the basis of manipulation of class-specific immunoglobulins has been designed in our laboratory (263). The immunoassay uses specific IgM and IgG capture and second antibodies, respectively. This allows use of a commercially available antimouse-IgG conjugated with alkaline phosphatase to specifically detect the viral antigen.

The attractive aspect of RIA and ELISA-based assays is that they allow both qualitative and quantitative measurements (Fig. 1) and are adapted for easy and rapid assay of multiple seed samples. This can be particularly important with seed assays, as demonstrated by investigations with detection of soybean mosaic virus in soybean seeds (65,66,116,233,311). The amount of soybean mosaic virus antigen that is determined in a soybean seed lot may, for example, allow estimation of secondary spread during a growing season (233) as well as estimation, on a quantitative basis, of the percentage of seed containing virus antigen (65). Of critical importance in such assays is the sampling method used and the limitation established for detection sensitivity. These aspects have been recently reviewed (180,324,325,490) and therefore are not addressed in this discussion.

It may be possible to electrophoretically separate proteins occurring in a crude protein extract of seed from selected plant species and, after probing a Western blot of the electrophoretic separation with antibody specific to viral coat protein, detect virus antigen. This procedure has been used to specifically

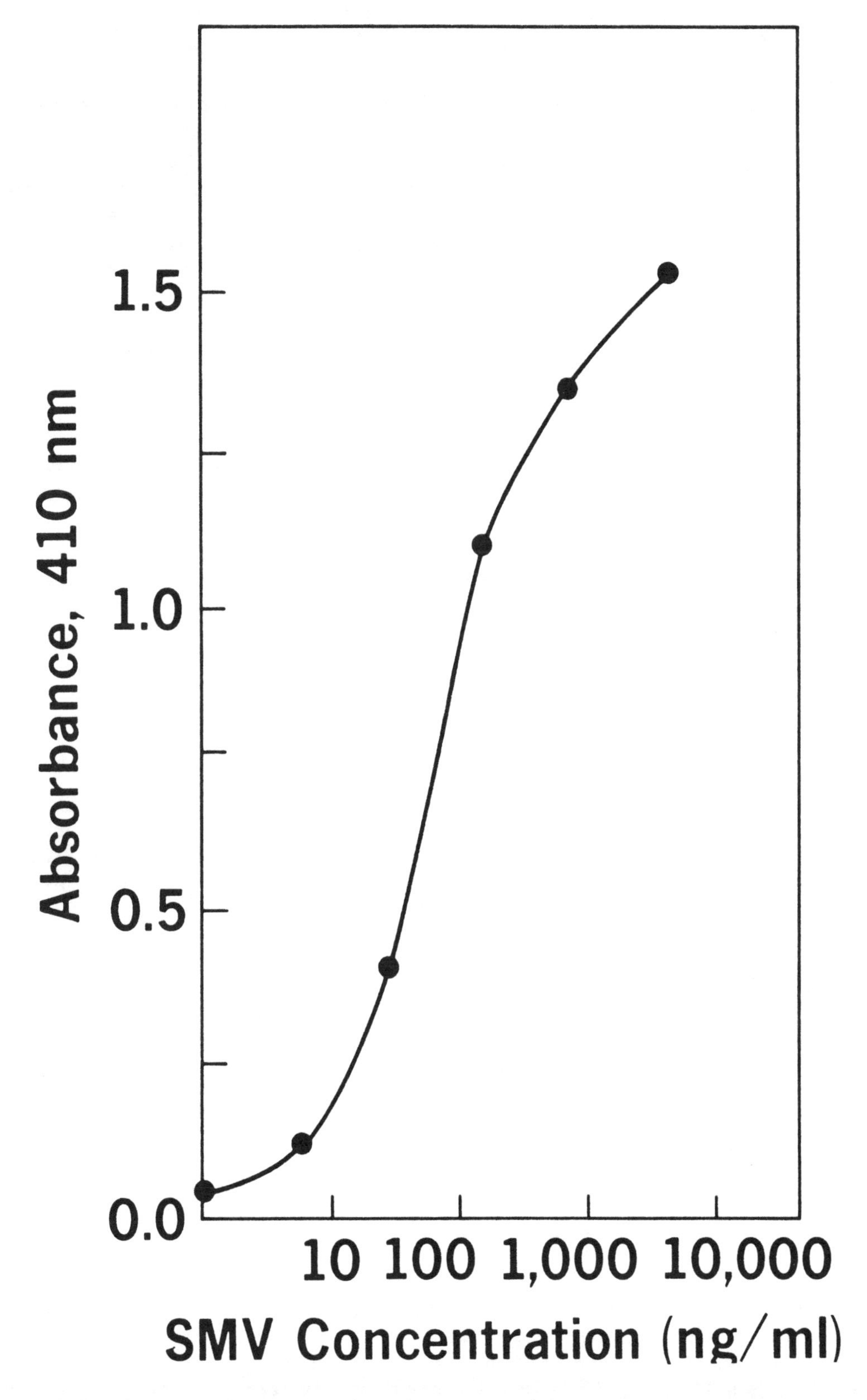

Fig. 1. Relationship between absorbance at 410 nm and concentration of purified soybean mosaic virus added to extracts of healthy soybean seed in a monoclonal antibody-based biotin-avidin ELISA.

detect soybean mosaic virus coat protein in a few infected soybean leaves (R. A. Andrews and J. H. Hill, unpublished), but it has not been attempted with seed and does not seem applicable to multiple sampling procedures. Similarly, serologic-specific electron microscopy employing polyclonal (63) and monoclonal (118) antibodies has been used to detect viruses in seed and leaf tissues but is not readily applicable to multiple sampling procedures.

An alternative approach to the detection of seed-borne pathogens is the demonstration of pathogen-related nucleic acid in plant tissue. The methods are generally not as extensively developed or utilized as those based upon serology. However, they do offer some potential and, particularly for detection of potato spindle tuber viroid, can be used effectively. The two approaches adopted are based upon electrophoresis and hybridization of nucleic acid with cDNA probes on nitrocellulose membranes.

Electrophoresis has been effectively used to detect potato spindle tuber viroid in potato seed pieces (390). The technique is based upon visualization of an RNA species that is present in diseased tissue but absent in tissue not infected with the viroid. Identification of the RNA species is based upon its comigration with known viroid RNA in polyacrylamide slab gels.

Although not used for identification of viruses in seeds, polyacrylamide gel electrophoresis has been used to detect double-stranded viral RNA species in infected leaf tissue and concomitant absence in uninfected leaf tissue (524). This method is perhaps most useful for putative detection of infection by any one of several viruses in a single virus family. It is based upon a nucleic acid banding pattern similar to those known to be characteristic of viruses belonging to discrete families. The potential efficacy of his technology for seed testing is unknown.

DNA probes are pieces of DNA that specifically recognize and bind to complementary sites on nucleic acid present in cells. Procedures for generation of cDNA probes from single-stranded positive-sense RNA plant viruses are well known and can be produced by using commercially available kits. Briefly, a cDNA copy is generated from a purified RNA of high quality by using the enzyme reverse transcriptase. The single-stranded RNA is digested away from the RNA-DNA hybrid by using alkali or S1 nuclease. The cDNA can then be labeled for use as a probe or, preferably, made double stranded by using DNA polymerase and be cloned into an appropriate plasmid. This will allow generation of a cDNA probe of consistent quality and sequence. After excision from the plasmid vector, the cDNA can be labeled with ^{32}P or biotin by nick translation (304). Labeling of the cDNA with ^{32}P yields a diagnostic probe with a short shelf life. In contrast, labeling with biotin provides a cDNA that can be used for a considerably longer time. Additionally, the recent procedure of photobiotinylation (168) makes biotin labeling of cDNA as well as antibody a very attractive option.

Detection of potato spindle tuber viroid in potato seed pieces is routinely accomplished by Northern blots using a ^{32}P labeled cDNA probe (379). In this application, nitrocellulose "dot-blotted" with extract of tissue to be tested is probed with a ^{32}P cloned cDNA to potato spindle tuber viroid. In a similar fashion, DNA from <u>Pseudomonas</u> <u>phaseolicola</u> is accurately detected in seeds of field-grown field beans by using a cloned cDNA to the bacterial toxin gene (441).

It is appropriate to recall that a major disadvantage of current molecular approaches to detection of plant pathogens is that detection of antigen or nucleic acid may include both infectious and noninfectious propagules of the pathogen. Therefore, although detection based upon biological and molecular approaches may not be absolute, time, labor, space, and economics may favor the utilization of molecular approaches. A further disadvantage of molecular probes, in contrast to certain biological tests, resides with limitations in detection sensitivity. Although the detection sensitivity of antibody and cDNA assays resides in the picogram and nanogram range, it is unlikely that very low levels of seed transmission, such as that shown for bean pod mottle virus in soybeans (310) and maize dwarf mosaic virus in corn (236), can be demonstrated by using these assays.

GRAIN QUALITY AND GRADING STANDARDS

David B. Sauer

Research Plant Pathologist, USDA, ARS,
U.S. Grain Marketing Research Laboratory
Manhattan, KS 66502

The quality of soybeans can be adversely affected by diseases in the field as well as by postharvest storage fungi. Infected or damaged beans may have poor viability and may be of lower quality for processing. When visible damage is present, the soybeans may be downgraded with a resultant lower market price. The principal causal agents and environmental factors involved in soybean damage will be discussed.

PREHARVEST DISEASES

Of the fungi that affect soybeans before harvest, the *Phomopsis-Diaporthe* group of pod and stem blight fungi are the most important. Infected seeds are smaller and less dense, split easier during handling, and have lower oil quality, poorer germinability, discoloration, and various taste and odor problems in the processed oil and flour (215,265,294,413). While these fungi may be widespread in immature beans (281), the levels of infection and damage increase greatly when the plants are mature and harvest is delayed (281,458,550). Many researchers have shown that rainfall or moisture after maturity is a key factor in determining the extent of damage by *Phomopsis* (265,430). Rupe and Ferriss (430) found a linear relationship between pod moisture and rate of infection from 35% moisture content down to 19%. No infection occurred when pod moisture was below 19%; above 35%, other organisms prevailed.

Studies that show higher disease severity in southern compared to northern locations (265) may be related to temperature differences. While both high humidity and high temperatures can favor disease development, Spilker et al. (479) found less infection at high humidity and low temperatures than at high humidity and high temperature. Early planting and early maturing varieties are associated with higher levels of *Phomopsis* seed damage (458,504,550). Several studies have shown this to be a function of exposure to certain weather conditions rather than to earliness or lateness per se. That is, late maturing plants simply escape seed decay problems (458), and, if late maturing varieties are forced to mature early, they have increased infection and reduced seed viability (549).

Other fungi, such as *Alternaria*, can also cause seed discoloration and poor viability, and they are also associated with wet, delayed harvest conditions (295). Insect damage can increase infection by *Alternaria* (461), but fungi have been shown to penetrate natural pores in soybean seed coats, indicating that insect damage or cracks are not required for infection (231).

"Weathering" is a term often used to describe the discoloration and damage from infection by late season fungi such as *Phomopsis* and *Alternaria*. Weathering was blamed for some recent major concerns about the quality of soybeans exported from the U.S. (327). Damaged beans were apparently blended with sound beans to create a mix with the maximum allowable percentage of damaged kernels for the particular grade. Standards allow a certain percentage of damage in each grade, but customers may routinely get shipments with considerably less than the allowable damage. When they suddenly receive beans with the maximum allowable level of a very conspicuous damage, they may think the grading system has changed or the shipment was improperly graded.

Care must be taken in grading soybeans to distinguish natural color variations from discolorations which are caused by disease and may be related to problems such as poor viability. For example, hilum color can be a varietal trait or it can be an indication of seed infection (367).

POSTHARVEST OR STORAGE DETERIORATION

Ramstad and Geddes (407) were the first to associate fungi with storage deterioration in soybeans, although in early studies of respiration and heating, it was not clear whether seed or fungal respiration was being measured. The indisputable role of fungi was demonstrated when heating was found to be similar in normal beans and in beans which were autoclaved and then inoculated (350). Subsequent work has elucidated the variety of effects caused by storage fungi and the environmental conditions which favor the different species.

Each species of storage mold has a minimum moisture content that is required for growth (96). If moisture is below a given level, we know that certain species cannot grow. At moisture contents of 13-15%, _Aspergillus glaucus_ and _A restrictus_ can grow (92,94,96,122,275) and will eventually produce mustiness, discoloration, etc. As moisture content increases, these fungi will grow faster and cause more damage. They will also be competing with other species that have higher minimum moisture requirements.

Most storage fungi are much more destructive than _A. glaucus._ When moisture is 15-16% or more, _A. candidus_ and _A. ochraceus_ can grow. _A. candidus_ is capable of rapid growth and is frequently involved in grain heating. It can grow at temperatures as high as 55 C. _A. ochraceus_ is capable of causing advanced spoilage and possibly mycotoxin production, but it is not commonly found in stored soybeans or other grains (96).

A number of _Penicillium_ species grow in seeds and grain, some at moisture levels as low as 16%. Some grow at temperatures near freezing or even below, but growth is slow at low temperatures. If grain with 18-20% moisture content is kept cool or cold, _Penicillium_ is likely to be the cause of spoilage (96,122). Some species can produce mycotoxins including ochratoxin, patulin, and tremorgens, but they have not been reported in soybeans.

Aspergillus flavus is probably the most notorious of the storage fungi, because of its ability to produce aflatoxin. Relative humidity must be above 85% (seed moisture above 18%) and temperatures fairly warm for _A. flavus_ to grow in soybeans. Under such conditions, it has the potential for rapid and destructive growth (42,350). It was the only fungus isolated from a large silo of soybeans that had undergone heating (42). In that case, no viable fungi were recovered from the center portion which had heated as high as 98 C, but _A. flavus_ was found in the adjacent areas.

As moisture increases, more and more fungi are capable of growing. Above 20%, some of the so-called field fungi such as _Fusarium_ may be able to compete, as can various yeasts and bacteria. At these high moistures, microbial activity is so great that respiration of the microorganisms will quickly change the interseed environment from aerobic to anaerobic. As oxygen decreases and carbon dioxide increases, the mycelial fungi are unable to compete with the anaerobic yeasts and bacteria.

Fungi respond to moisture availability or water activity in their environment. In grains and seeds, we express moisture as the percent of water in the seeds. If 100 g of grain contains 15 g of water, we say the grain has a moisture content of 15%. The water activity or relative humidity that is in equilibrium with 15% moisture is different for different seeds. Conversely, different kinds of seeds will have different moisture contents when they are in equilibrium with 85% relative humidity. These relationships are shown in Figure 1 for corn, wheat, and soybeans at 25 C (D. B. Sauer, unpublished). The curves for wheat and corn, typical of starchy cereal grains, are quite different from soybeans. At low relative humidities, soybeans have moisture contents lower than the others, but in the biologically important range of 75-85% relative humidity, soybeans equal, then exceed, corn and wheat. Because of this complex relationship, we cannot generalize that, because of moisture, fungi grow faster on soybeans than on corn at equal moisture contents. The generalization may be true at low moisture contents but not at high moisture contents.

Temperature in storage also has a major effect on the growth of fungi. Christensen (92) found little fungal growth in soybeans stored 16 months at 14% moisture and 5 C, but at 25 C invasion by A. glaucus and A. restrictus was extensive within 5 months. By 16 months at 25 C, fat acidity values had increased greatly. Damage in storage at 20 C has been shown to be much less severe than at 25 C (94). Deterioration is slower at lower temperatures, even at high moisture contents. Beans stored at 18-19% moisture and 15 C were invaded by A. glaucus after 6 months, but germination was still high (122). In other work, fungal invasion, fat acidity increase, and germination decrease were rapid at 30 C, but at 10 C beans remained in good condition for 22 weeks at moisture contents up to 17.3%.

Two of the most common parameters measured along with fungal invasion are germination and fat acidity. Several studies have shown that germination can decrease in soybeans even when stored under good conditions without fungal growth (42,70,94). Germination as well as various other seed quality factors decrease gradually in storage (70). Different lots of beans vary widely in their ability to maintain seed viability, but, in general, germination seems to be less stable in soybeans than in cereal grains.

Baker (32) found that fat acidity correlated with both mold damage and heat damage in graded soybean samples. Other workers have documented the increase in fat acidity with increasing time in storage under conditions suitable for mold growth (92,94,122,178,436). In addition to increase in free fatty acids, other lipid changes accompany fungal invasion, especially when the spoilage is severe (351,413). Cereal grains sometimes become discolored or brown under conditions of mold growth and heating, but soybeans are much more susceptible to browning. The browning can apparently be attributed to the production of phenolic acids by fungi growing in the beans (172), although earlier work implicated a Maillard-type sugar-protein interaction (351).

Heat is produced by both the respiration and browning reactions in soybeans, and cases of severe heating are not uncommon (42). Navarro et al. (362) reported that heating in a bin of soybeans with 13.9% moisture was stimulated by warm summer temperatures and an accumulation of fine material in the center of the bin. Aeration and refrigeration have been used to control or prevent heating, but in many cases when heating is first detected the bin is emptied and the beans are processed immediately (42,350). The greater susceptibility of fines and broken kernels to fungal growth and heating has been recognized in all kinds of stored grains and seeds. Ramstad and Geddes (407) found that split beans respired 6 to 24 times faster than whole beans.

If heating is not arrested, soybeans will become dark brown to black with a carbonized appearance. This has led elevator operators to file claims of fire damage in an attempt to recover from an insurance company some of the losses which were caused by their own negligence. However, it is possible to distinguish between naturally heated or binburned soybeans and those which have actually been fireburned (95,351). Such evidence is used in courts of law to settle damage claims.

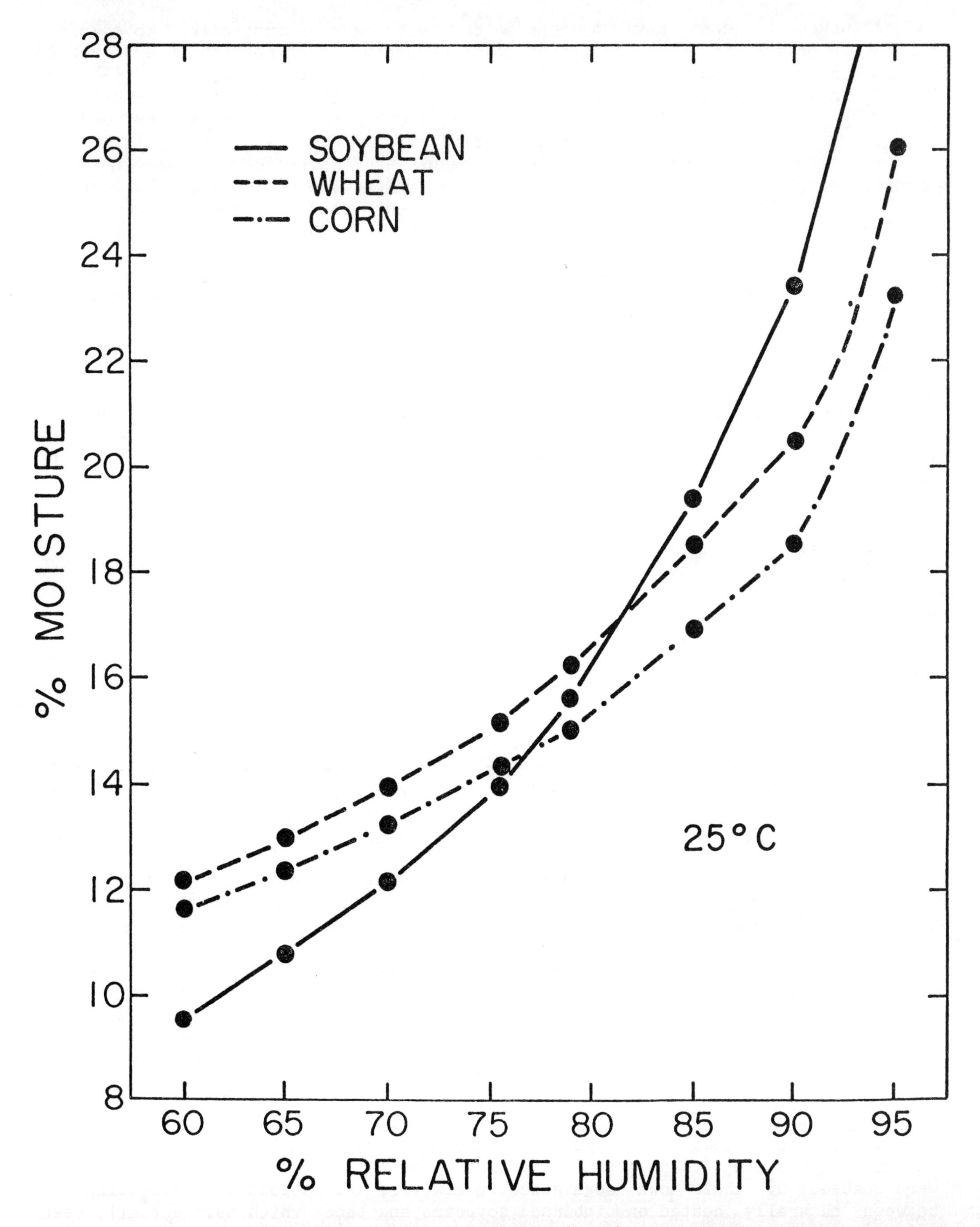

Fig. 1. Adsorption moisture content of soybeans, wheat, and corn in equilibrium with relative humidities of 60-95% at 25 C.

Although soybeans have a reputation as being more susceptible to mold and heating problems in storage compared to other grains, they have not been shown to be a high-risk crop with regard to mycotoxins. Shotwell et al. (462,463) have tested many commercial samples of soybeans and other grains for mycotoxins including aflatoxin. They found two of 1,046 soybean samples to contain aflatoxin; both were in U.S. sample grade, the lowest grade. Bean et al. (41) reported aflatoxin contamination in soybeans but Shotwell et al. (462) found no contamination even in a group of 180 samples which were mostly from the southern U.S. Attempts to produce aflatoxin on various substrates have shown poorer yields on soybeans compared to other grains and oilseeds (464). Phytate and zinc have been implicated as factors which may limit aflatoxin production in soybeans (211), and autoclaving the beans increases the amount of aflatoxin which can be produced. The factors which make soybeans more mold resistant or less suitable for aflatoxin production may vary widely among varieties (464).

There does not appear to be a simple answer to the problem of whether moldy soybeans or other moldy materials are harmful to animals or humans who consume them. Even in the absence of mycotoxins, we may question the nutritive value or wholesomeness of moldy grains. While some tests may show poorer performance of animals fed moldy diets, it is possible to show not only satisfactory, but improved performance on materials heavily invaded by fungi. Chah et al. (87) grew various fungi on soybeans and used them in diets for chicks. Three species of *Aspergillus* produced faster growth and increased feed efficiency compared to normal soybeans.

The effects of storage damage on the use of soybeans for oriental food products such as tofu and soy milk have been studied. In addition to increases in fat acidity and discoloration, Saio et al. (436) measured decreases in protein extractability, greatly increased leakage of solutes into soakwater, and a deterioration of several characteristics of tofu gels. All of these deteriorative changes increased with storage time and were greater as storage temperature or humidity increased.

SOYBEAN GRADING

The main factors and requirements for U.S. grades of soybeans are shown in Table 1 (523). Each factor in Table 1 can be affected by diseases or storage fungi, although the damaged kernel factor is the category most likely to be limited by fungal problems. Reduction in test weight is mainly associated with diseases in the field, and heat damage is strictly a product of storage conditions. Although the grading procedures assume the inspector can distinguish between damage from storage molds and field molds, and damage and odors caused by fire compared to biological and chemical heating, it seems likely that such distinctions at times may be unclear.

A condition referred to as "purple mottled or stained" is a special category and is addressed as follows in the grading handbook (523): "Purple-mottled soybeans are soybeans which have their seed coat discolored pink or purple. This discoloration is the result of a fungus growth and may cover all or part of the kernel. . . . A distinctly discolored soybean shall be a soybean which has one-half or more of its surface area covered by the growth of a fungus; . . . Soybeans which contain 2.0% or more of distinctly discolored soybeans shall be considered Purple Mottled or Stained. . . . this fact shall be shown on the pan ticket and the official inspection certificate and the soybeans shall not be graded higher than U.S. No. 3."

In grading damaged kernels, the damage must be distinct. "In general, a soybean shall be considered to be damaged for inspection and grading purposes only when the damage is distinctly apparent and of such character as to be recognized as damaged for commercial purposes." Each category of damage is described in the

Table 1. Grades and grade requirements for soybeans.

Grade	Minimum test wt. per bushel	Maximum limits of			
		Splits	Foreign material	Heat damaged kernels	Damaged kernels total
	lb	%	%	%	%
No. 1	56	10	1	0.2	2
No. 2	54	20	2	0.5	3
No. 3	52	30	3	1.0	5
No. 4	49	40	5	3.0	8
Sample	(does not meet above requirements or is musty, sour, heating, contains stones, etc.)				

handbook and also in a set of color slides called interpretive line slides. If the level of damage on a kernel in question equal or exceeds that on the appropriate line slide, it is graded as damaged; otherwise it is not.

Heat damage is treated more severely, as indicated in Table 1, a reflection of the extensive chemical deterioration that accompanies it. Designation of heat damage is complicated in soybeans by a special category in the instructions to inspectors: "Slight Discoloration by Heat. Soybeans and pieces of soybeans which have been damaged by heat, but which are not materially discolored, shall be considered damaged, but not heat damaged."

Soybean processors do not like to use beans that have a high free fatty acid content. They have advised the Federal Grain Inspection Service (FGIS) that grade 2 soybeans should not exceed 0.75% free fatty acids, but that level is often exceeded in samples with the maximum 3% damaged kernels. In September 1986, FGIS began using a new set of interpretive line slides for soybean damage factors to address the problem. The new slides will result in more moderately damaged kernels being graded as damaged. Using samples graded by the old and new criteria, FGIS (255) showed that free fatty acid content was lower using the new slides. For example, samples with 2% and 3% damage had average free fatty acid contents of 0.90 and 0.97%, respectively, with the old interpretations and 0.67 and 0.71% with the new.

Market prices for soybeans are usually quoted for U.S. No. 1 grade but not for lower grades. In actual practice, buyers will specify discounts for individual factors of concern. The discounts may be quoted as a percentage or as cents per bushel. Excess moisture above a certain level, such as 13%, will be penalized a percentage of the weight, and the rate of penalty will increase at higher moisture contents. Soybean processors may set strict limits for various factors in addition to the discount schedule. One buyer, for example, will not at present accept shipments that exceed 15% moisture, or 8% total damage, or 5% heat damage, or that are less than 50 lb/bu.

A local elevator quoted the following discounts for soybeans: moisture content 13% or less, no discount; 1% reduction for each 0.5% moisture above 13.0%; 2% reduction for each 0.5% moisture above 15.0%. Test weight at least 54 lbs, no discount; below 54 lbs, 1/2 cent per bushel for each lb. Damaged kernels no more than 2%, no discount; then $.04 per bushel discount for each percent damage. Heat damage is discounted $.04 per percentage point. Full market price would be paid for soybeans that meet all grade requirements for No. 1 except test weight, which must meet grade No. 2, and the beans cannot exceed 13% moisture content. At $5.00 per bushel, beans with 14% moisture would be discounted $.10 per bushel. Beans with 15% damage as the only discount factor would bring $4.48 per bushel, a 10.4% price reduction.

Discounts for various grade factors vary from year to year and with geographic area depending on supply and demand, and depending on the extent any given factor is a grading problem in that area. However, competition among elevators and processors tends to keep discount schedules fairly uniform within a local area.

MIDSEASON SOYBEAN DISEASES

Bill Kennedy

Department of Plant Pathology
University of Minnesota
St. Paul, MN

Significant midseason diseases include a variety of bacterial, viral, and fungal pathogens. Origin and destiny of these diseases overlaps with the seed/seedling diseases and the "late season" diseases. Nevertheless, these midseason types represent the major problem diseases of soybean, and their control is vital to cost-effective production. Approaches to management include genetic, cultural, and chemical methods; application of each is a matter of judgment and the role of the plant pathologist as a principal coordinator among colleagues cooperating in the industrial and academic sector is essential.

We know much about some of these important problems, but research efforts must proceed within a continuous, long-range time frame. In the case of brown stem rot, crop rotation, though not practical in some cases, can effect control of this important and widespread disease in the midwest (137). The disease caused by *Phytophthora*, on the other hand, requires a continuing genetic manipulation in order to avoid vulnerability of the host via appearance of new races of the fungus.

Other diseases, such as "Sudden Death Syndrome," require fundamental study on cause and effect; the epidemiological perspective on such a disease remains largely unknown. Controlling the destructive cyst nematode requires a variety of approaches on fundamental understanding of population genetics and ecology, as well as practical cultural practices in changing agricultural land use patterns. Stem canker, until recently a disease of the past, has become more important and is receiving attention dictated by its rise to prominence in important soybean producing regions of the south. The problems caused by *Sclerotinia*, normally a weak or non-aggressive pathogen to soybeans in most areas of production in the United States, require increased attention as types and prevalence of cropping sequences increase.

Responsible professionals must be aware of both neglect and scare tactics in disease assessment. Our facility for measuring loss from disease attack is incomplete and has far-reaching effects on economy of production and distribution of resources for research. We have very little useful information on loss from bacterial and viral diseases. The varied fungal leafspots, *Phytophthora*, pod and stem blight, and nematode disease, where chemicals can be used for loss studies, are complicated by both beneficial and detrimental chemical effects as well as those related to the disease under study. It is germane to reflect on the support provided and projected for studies on midseason diseases. There is a general decline in funds for production research nationwide, and this fact will reflect on the kinds of things we do to provide our clientele with the optimum mix of disease management strategies.

BACTERIAL, FUNGAL, AND VIRAL DISEASES AFFECTING SOYBEAN LEAVES

J. M. Dunleavy

Research Plant Pathologist
Agricultural Research Service
U.S. Department of Agriculture
Department of Plant Pathology
Iowa State University
Ames, IA 50011

Because many of the constituents needed by soybean plants for seed formation depend on leaf photosynthesis, and because disruption of the photosynthetic process by disease-producing microorganisms may result in a reduction of seed size, seed number, or both, diseases affecting soybean leaves have been the focus of plant pathologists for many years. Ratings of soybean disease monitoring plots at nine locations in Illinois revealed that brown spot, bacterial blight, bacterial pustule, and downy mildew were the major foliar diseases in 1980 (264). Maximum percentage of leaf area of susceptible cultivars infected were: bacterial blight, 16%; bacterial pustule, 28%; brown spot, 33%; and downy mildew 17%. Other diseases affecting leaves, such as Cercospora leaf spot, are important in Arkansas and other southern states. My objective is to present a few of the salient symptoms and recent research progress made in studies of some of the major mid-season diseases affecting soybean leaves.

BACTERIAL LEAF SPOTS

Bacterial Blight

Bacterial blight, caused by *Pseudomonas syringae* pv. *glycinea* (Coeper) Young, Dye, and Wilkie, is one of the most widespread soybean diseases. Bacteria splashed from soil or nearby diseased plants invade leaves through stomata and multiply in intercellular spaces of the mesophyll, where they produce a toxin that inhibits chlorophyll synthesis (123,199). The bacteria multiply rapidly and produce small, angular, wet spots in about 7 days. The bacteria can multiply on leaf surfaces of healthy soybeans and remain viable without causing infection. Early in the growing season, pathogenic races of the bacterium tend to be specific to the cultivar on which they occur. Late in the season, bacteria pathogenic to several cultivars tend to increase on many cultivars(348). If the host is susceptible, the lesions become chlorotic in 3-to-5 days, and subsequently necrotic.

Distribution of diseased plants in nine fields near Ottawa, Canada, was determined. Diseased plants were distributed nonrandomly from early to midseason. Samples were taken to estimate disease incidence. Simple random samples were inadequate to assess disease percentage when the underlying distribution was nonrandom (401).

Bacterial blight is most severe during periods of cool weather and frequent rain or dew. It is usually observed early in the season before plants flower (382). When a leaf has many infection sites, lesions coalesce to form large necrotic areas that may fall out. Such leaves may be severely torn by high winds during rainstorms. Soybean bacterial tan spot produces similar symptoms, but the two diseases can usually be distinguished by presence of angular wet spots on adjacent leaf tissue when bacterial blight is present. Tan spot lesions never have a wet appearance.

A suspension containing 10^8 colony forming units (CFU) of the causal bacterium per ml was sprayed into a chamber containing susceptible soybean seedlings. An epiphytic population of 10 CFU per ml developed, but no infection, in 2 weeks on leaves that were wetted gently with atomized water before inoculation and kept at 20 C without further wetting. Bacterial blight developed on leaves that were either wounded and wetted, or water-soaked and wetted immediately before inoculation (485). Once seedlings were infected, the injury produced by the disease could be detected by cotyledon chlorophyll concentration (277).

After glyceollin was reported to be related to the expression of bacterial blight resistance in soybean leaves to incompatible bacteria (274), it was observed that inhibition of glyceollin accumulation by glyphosate did not result in a complete reversal of plant reaction from incompatible to compatible (247). Other workers (165) reported that glyceollin had no inhibitory effects on blight bacteria when six different bioassays were used.

The role of bacterial immobilization in race-specific resistance of soybean to blight bacteria was investigated, and bacterial envelopment was not observed, even though a typical hypersensitive response was induced, indicating that bacterial attachment per se is not a prerequisite for induction of the hypersensitive response in soybean (164). Because of the diversity of races of blight bacteria, development of resistant cultivars is not a practical control measure.

Bacterial Pustule

Bacterial pustule, caused by the bacterium Xanthomonas campestris pv. phaseoli (Smith) Dye, is present to some extent in most soybean-growing areas. In the North its prevalence and severity seem to vary considerably with the season, but in the South it is more uniformly severe when susceptible cultivars are grown. Lesions produced by the bacteria usually are confined to leaves. At first they appear as small, yellowish-green areas with reddish-brown centers, more conspicuous on the upper surface of the leaf. A small, raised pustule usually develops at the center of the lesion, especially on the lower leaf surface. This is the stage at which the disease is most readily distinguished from bacterial blight. The pustule and the absence of a wet appearance before the lesion turns chlorotic distinguish bacterial pustule from bacterial blight.

When dispersal of pustule bacteria was studied on soybean hill plots planted 0.76 m apart, the pathogen was dispersed equally in all directions and at the same rate (0.1 m/day) in plots of resistant and susceptible soybeans. Populations of the pathogen on resistant and susceptible cultivars were not significantly different during the first 3 weeks of the study, and peak populations occurred at 35 days when populations were 40-fold higher on susceptible plants (196).

In a study to determine the effects of sulfur dioxide on expansion of lesions caused by pustule bacteria in soybean, lesion development was inhibited and time of exposure altered the subsequent effect on lesion size (299). In another test, the design resulted in exposures to sulfur dioxide on the first 2 days of the week, inoculation on the third day and exposures on the fourth and fifth days; at the conclusion of the test, plants exposed on days 1, 2, or 3 consistently developed smaller lesions than those exposed on other days (298).

The most effective control measure for bacterial pustule is use of resistant cultivars.

Bacterial Tan Spot

Bacterial tan spot, caused by Curtobacterium flaccumfaciens (Hedges) Collins and Jones, is the only gram^{+} bacterium infecting soybeans. Leaf chlorosis usually is the first macroscopic symptom observed on naturally-infected field plants. Lesions frequently begin at a leaflet edge and progress inward toward the midrib. After several days, the chlorotic tissue usually dries and assumes a tan color. Commonly, a single lesion on a young expanding leaflet may spread over an entire leaflet. Such leaves eventually fall from the plant. If leaves are older when infected, lesion size is usually smaller. Necrotic areas of leaflets usually fall from the plant during high winds, giving a ragged appearance to the leaves (128). The disease was originally reported from Iowa, and in 1982 it was found in 51% of 818 fields examined (136). The disease was localized in fields, and infested portions varied from a single plant to areas up to 10x30 m. These elongate diseased areas were oriented with the longest dimension in the same direction as the rows. The disease was not believed to be a significant problem in any of the fields observed.

A study to determine if tan spot occurs in other soybean-producing states was conducted in cooperation with soybean breeders from 17 states, each supplying seed samples of cultivars grown in the Uniform Soybean Tests in 1985. The pathogen was found at 75% of 44 locations sampled and was distributed among 13 states (includes Province of Ontario, Canada). It occurred as far south as Texas and as far east as Maryland (134).

Bacterial tan spot is easily distinguished from other bacterial diseases of soybean, but the pathogen produces symptoms similar to Phyllosticta leaf spot, a fungus disease. The only certain method of distinguishing the two diseases is to isolate and test for the causal organism.

Disease reaction of 20 cultivars grown in the northern United States (maturity groups I-IV) were as follows: resistant, three; slightly susceptible, seven; susceptible, three; and very susceptible, seven. Twenty cultivars grown in the southern United States (maturity groups V-VIII) were more resistant as a group: eight were resistant and 12 were slightly susceptible (128).

C. flaccumfaciens is seed transmitted in soybeans. When 10 seeds from each of 50 field-grown plants of a susceptible cultivar were grown in the greenhouse, tan spot symptoms appeared on 26% of the seedlings. Of the original 50 plants tested, 90% transmitted the bacterium to some seedling progeny. The mean bacterium transmission per plant was 2.6%, and the range was 0-90% (130).

Temperature has an effect on systemic spread of the pathogen from seeds to unifolioliate leaves. When seeds exposed to natural infection were germinated at 10-35 C, the bacterium infected leaves only at 25 and 30 C (133). It is from such leaf lesions that the pathogen spreads to other plants in the field during the growing season.

The spatial distribution of tan spot was observed over time in four field plots in which two soybean plants were inoculated near the center of each plot. Mean size of the four areas of diseased plants 5 weeks after inoculation was 19x5 m (132).

In a study conducted at three locations in Iowa, mean yield losses due to tan spot were 13% in 1979, no loss in 1980, and 4% in 1981. An 8% loss was measured at one location in 1978. Seed yield losses in susceptible cultivars ranged from 0 to 19% (129).

Cupric hydroxide was effective as a protectant in the field against C. flaccumfaciens in uninoculated, susceptible soybeans. It failed, however, to eradicate the disease in inoculated plants (129).

FUNGUS LEAF SPOTS

Brown Spot

Brown spot, caused by the fungus *Septoria glycines* Hemmi, was formerly regarded as a disease of minor importance but has increased in prevalence in the Midwest in recent years. It appears early in the growing season on primary leaves of young soybean plants as angular, reddish-brown lesions that vary from the size of pinpoints to 3 mm wide. As the plants grow, the fungus spores produced on the primary leaves spread and infect trifoliolate leaves, stems, and pods. Heavily infected leaves gradually turn yellow and fall prematurely. Defoliation progresses from the base toward the top of the plant. In severe cases the lower half of the stems may be bare of leaves before maturity.

There is little information available on the effect of dew on diseases caused by pathogens such as *S. glycines* whose conidia are disseminated by splashing rain. In a study in which misting soybean foliage at night to simulate dew formation was used, daily simulated dews during the infection period (44 hr) caused increased infections on plants receiving the treatment compared to controls (418). In a study of the influence of overhead irrigation and row widths of 15, 45, and 90 cm on brown spot development, there was no increase in the disease due to the treatments (239). A minimum leaf wetness period of 72 hr between 16 and 32 C, with optimum of 28 C, was required for brown spot development (389).

Bright field and electron microscopy revealed that *S. glycines* may require up to 72 hrs to enter lower leaf stomata. By day 5 palisade cells collapse and become necrotic, even though the hyphae are in the spongy mesophyll. Maximum necrosis is reached by day 14, and pycnidia are formed during the 3rd week (138).

Field evaluation of 15 isolates of *S. glycines* from 12 states was conducted to determine if isolates varied for pathogenicity. Results showed that pathogenic variability among isolates was not detectable in the field (266).

There have been several studies of yield losses caused by brown spot. Infected Wells soybeans yielded less than uninfected controls, and seed weight was reduced as well. The rate of leaf loss was greater for infected plants during pod-fill, and the rate of dry matter accumulation in pods of infected plants was less (194). In yet another study, seed weight reductions in the upper, middle, and lower canopy levels were 8, 11, and 16%, respectively (384). In yet another study, yield reductions among inoculated plants ranged from 8 to 11% in 1978, and from 12 to 14% in 1979 (308). Benomyl was an effective fungicide for control of brown spot (385).

Frogeye Leaf Spot

Frogeye leaf spot, primarily a disease of leaves, is caused by *Cercospora sojina* Hara. When the fungus infects the leaf tissues, it causes an eyespot type lesion with gray or light-tan central areas and narrow, reddish-brown borders. Heavily spotted leaves fall prematurely. Infection also can occur on stems and pods late in the growing season. The fungus sometimes enters the seed from infected pods.

Small lesions are considered a resistant reaction and large lesions a susceptible reaction but time-lapse photography of lesion development showed that lesions did not enlarge on over 70 cultivars inoculated with five races of *C. sojina* (391).

Kent was found to be a mixture of reaction types to race 5 of *C. sojina* in six of seven seed sources tested, and to race 2 in two of seven sources. Kent was considered resistant to race 2 and susceptible to race 5. The Rcs gene in Kent for resistance to race 2 cannot condition resistance to race 5, and a single dominant gene in Davis for resistance to races 2 and 5 was assigned the symbol, Rcs (52).

Downy Mildew

Downy mildew, caused by Peronospora manshurica (Naum.) Syd. ex Gaum., is one of the most common soybean diseases in the United States. The fungus produces chlorotic areas from flecks less than 1 mm wide up to lesions 1 cm wide. During periods of dew, conidiophores form on the lower surface of infected leaves. Conidia spread the disease from plant to plant. After conidial production, oospores form in internal portions of lesions, after which the lesions become necrotic. The fungus grows systemically in susceptible plants, invading pods and covering seeds with a white crust of oospores.

Resistance to downy mildew is conditioned by the gene Rmp which is included in the genome of Union. Prior to 1981, Union was resistant to all known races of P. manshurica, but in that year, Union was infected by a new biotype of the fungus. This biotype has been reported only from Illinois (309).

Leaves of Union are resistant to race 24, but leaves of Williams are susceptible. Union and Williams are nearly isogenic cultivars. Union, when inoculated, reacted with a typical hypersensitive response (147). Necrotic host cell death occurred around infection sites in Union based on epifluorescence observations of sodium fluorescein uptake, while Williams produced typical symptoms on leaves.

Information on the prevalence of downy mildew in Iowa was obtained in 1982, when the disease occurred in 50% of 825 soybean fields examined. The disease was observed in 85% of 95 counties sampled. Greatest prevalence of downy mildew occurred in an 11-county area in the north central section of the state where 89% of the fields were affected. Mean disease severity for all affected fields was classified as moderate [a rating of 2 on a scale of 1 (slight disease) to 4 (very severe disease)] (135).

Susceptible soybean leaves inoculated with conidia of P. manshurica developed leaf chlorosis in 8.5 days at 10 C and in 4.2 days at 35 C. A high negative correlation was recorded between air temperature and the appearance of chlorosis. Infection resulted in large increases in peroxidase at all temperatures tested between 10 and 35 C. The rapid increase in enzyme activity (a 967% increase at 25 C) in Peronospora - infected leaves may be a plant response to production of peroxide, or hydrogen donors, by the fungus, since peroxidase is an inducible enzyme (127).

Soybean yield losses caused by P manshurica were determined in the field in Iowa in a 2-year study. Metalaxyl-sprayed susceptible cultivars Wayne, Woodworth, and Williams 79 yielded 14, 9, and 8% more, respectively, than the same nonsprayed cultivars in 1983, and 18, 9, and 12% more in 1984. The combined mean loss of these cultivars, caused by downy mildew, was 10% in 1983 and 13% in 1984 (131).

Cercospora Leaf Spot

Cercospora kikuchii (T. Matsu & Tomoyasu) Gardner causes Cercospora leaf spot and purple seed stain; however, some isolates from seeds do not cause leaf spot symptoms. Symptoms first appear on leaves from the beginning through full seed set. Reddish purple, angular-to-irregular lesions occur on both upper and lower leaf surfaces. Lesions vary from pinpoint spots to irregular areas up to 1 cm wide and may coalesce to form large necrotic areas. Veinal necrosis also may occur. Infection of leaves and petioles causes rapid defoliation beginning with the uppermost leaves and moves down. Resistance varies among cultivars. Temperatures of 28-30 C with extended periods of high humidity favor disease development. The disease is more severe on early-maturing cultivars than on those that mature later at lower temperatures (535).

Soybean seedlings inoculated with *C. kikuchii* and placed in a dew chamber for 24, 48, or 72 hr at temperatures ranging from 16 to 36 C showed maximum infection when plants were exposed to a 24 hr dew period at 20 to 24 C. These results establish procedures to test for resistance to *C. kikuchii* (322).

The influence of overhead irrigation and row widths of 15, 45, and 90 cm on Cercospora leaf spot severity was studied in Arkansas. The disease was severe at all row widths tested, but developed more slowly at narrow row widths (239).

VIRAL DISEASES

Soybean Mosaic

Soybean mosaic is caused by the soybean mosaic virus (SMV). Seedlings from mosaic-infected seeds are short and spindly, with leaves that sometimes are mottled or rugose and that turn down at the margins. Field plants infected early in the season may be stunted, with shortened petioles and internodes. Leaves are reduced and may be asymmetric and puckered. Cultivars Hill and Essex and two introductions of *Glycine soja* were studied in the greenhouse for their response to the disease. Nodulated plants of the two *G. soja* introductions, as well as plants of Essex, infected with the virus, had significantly higher total N than uninoculated controls, and leaves developed higher virus concentrations than did nodules (377).

In a study of the ultrastructural cytology of soybeans infected with mild and severe strains of SMV, the cultivar Essex gave a tolerant reaction when infected with mild strains of the virus. A different strain isolated in Virginia induced severe mosaic symptoms and abundant pinwheel inclusions in the cytoplasm of infected cells. Cells infected by the mild strains produced both pinwheels and cytoplasmic strands. Cytoplasmic strands contained virus particles and traversed the vacuole of infected cells. Production of cytoplasmic strands was concluded to be an intracellular virus localization mechanism leading to the tolerant reaction of Essex soybean to the mild strains of the virus (250).

Two isolates of SMV were collected in Virginia and characterized. One isolate, SMV-VA, was classified into the G1 strain group. It infected only soybean cultivars known to be susceptible to SMV, and also infected cowpea and Alaska pea systemically, and Topcrop bean locally. The second isolate, SMV-OCM, caused severe systemic necrosis on SMV-resistant Marshall and Ogden soybeans, reactions typical of the G3 strain group (249).

A new source of resistance to the seven strains of SMV was found in Suweon 97 soybean obtained from Korea. Reactions of this line to each of the strains were symptomless in greenhouse tests (307).

Bean Pod Mottle

Leaf symptoms of soybean plants infected by the bean pod mottle virus (BPMV) are most easily observed near the plant apex where young leaves exhibit green to yellow mottling. Leaf symptoms are most obvious during periods of rapid plant growth under cool conditions. BPMV is sap-transmissible and readily transmitted by bean leaf beetles. When evaluating soybean lines for BPMV resistance using beetle vectors, researchers should group lines according to height to reduce variability in BPMV transmission associated with row-to-row variation in plant height. There is a high correlation between height of lines and disease incidence or symptom severity (554).

Soybean callus protoplasts have been infected with BPMV, but the success of infection is highly dependent on the pH of the K-phosphate buffer used. At pH 5.6, the optimum pH for infection without amendment, the addition of calcium chloride further increased infection. The stimulatory effects at pH 5.6 occurred only when the virus was exposed to the buffer prior to inoculation of the protoplasts, indicating that the buffer effects are primarily on the virus rather than the cells (305).

BPMV was not known to be seed-transmitted before 1983, but in that year the disease was reported to have been found in Nebraska for the first time and was observed to be seed transmitted (310).

BPMV may be found in doubly-infected plants in association with SMV. The combination causes severe losses: BPMV alone may reduce yields 10-15%, but in association with SMV, losses may be as great as 60%. Maximum losses occur when plants are infected at the seedling stage. The concentration of BPMV in doubly-infected plants was significantly higher than in singly-infected plants (71).

Bud Blight

Bud blight is caused by the tobacco ringspot virus (TRSV) and results in severe stunting when young plants are infected. Stunting is not evident in infected greenhouse-grown plants above 25 C. The most striking symptom is the curving of the terminal bud to form a crook. As the disease progresses, other buds on the plant turn brown and are brittle. The virus has been found in roots of seedlings after mechanical inoculation of primary leaves (321).

Examination of 630 plant introductions of *Glycine soja* in a search for resistance to TRSV resulted in finding resistance in one introduction, PI407287 (376).

Bud blight symptoms caused by other viruses infecting soybeans have been reported. Brazilian bud blight caused by the tobacco streak virus cannot be distinguished from bud blight on symptoms only, and similarly a soybean disease caused by the cowpea severe mosaic virus produces typical bud blight symptoms in both the United States and Brazil (19).

SOYBEAN STEM CANKER: AN OVERVIEW

Elisa F. Smith and P. A. Backman

Department of Plant Pathology
Auburn University

Stem canker was first recognized in the northern United States as a symptom variation of pod and stem blight, a disease of soybean (Glycine max (L.) Merr.) caused by Diaporthe phaseolorum (Cke. & Ell.) Sacc. var. sojae Wehm. (Dps). Dps and its anamorph Phomopsis sojae were reported to attack plants during senescence and were found to be responsible for reductions in seed quality (281,449). In 1948, Welch and Gilman (541) reported two distinct types of Diaporthe on soybean, a homothallic and a heterothallic type. The homothallic type was an aggressive pathogen on soybean causing cankers and girdling on actively growing plants. Welch and Gilman found substantial morphological and pathogenic differences and identified the organism associated with cankers as Diaporthe phaseolorum var. batatis. A comparative study of fungi causing stem canker and pod and stem blight reported in 1954 by Athow and Caldwell (22) confirmed the differences and separated these pathogens into two distinct varieties with characteristic morphology and pathogenicity. Further, they redescribed the causal organism of stem canker and defined it as a new variety of Diaporthe phaseolorum under the trinomial Diaporthe phaseolorum var. caulivora (Ath. & Cald.) (Dpc).

DISTRIBUTION

Beginning with the first published report of stem canker in Iowa in the late 1940's, the disease spread rapidly and became prevalent in the upper Midwest and Canada by the early 1950's. Stem canker was considered one of the most destructive diseases of soybean attacking actively growing plants and reducing yields sometimes as much as 40-50%. Symptoms of northern stem canker appeared late in the season (>70 days post plant), first as reddish-brown lesions on main stems, usually at the nodes or near leaf scars. As the disease developed, the lesion elongated several inches and became a definite sunken canker which could completely girdle the stem. Plants infected by Dpc prematurely died and retained their dead leaves. The impact of stem canker in the north was diminished when the highly susceptible cultivars Hawkeye and Blackhawk were eliminated from production fields. However, the frequent reports of isolations of Dpc from seed (281,449) suggest that the pathogen is still endemic in the upper Midwest.

In 1973, stem canker was found for the first time in the South (26). The disease first appeared in fields in Mississippi followed by reports in Alabama in 1977, Tennessee in 1981, South Carolina and Georgia in 1982, Florida, Louisiana, and Arkansas in 1983, and Texas in 1984 (28). Stem canker of southern soybean is symptomatically similar to the disease described by Athow and Caldwell in the North in 1954. However, differences have been observed in pathogenicity, etiology, and symptom expression. Cankers in southern-grown soybean are usually much more delimited and unilateral in appearance, progressing acropetally in infected stems. In addition to the differences in canker appearance, multiple stem infections are common with southern stem canker disease but are rarely observed with the disease in the North (28). In contrast to the northern isolates, the southern stem canker isolates are more aggressive, and attack a wide range of susceptible cultivars and caused an estimated $37 million in losses in southeastern United States in 1983.

TAXONOMY

Athow and Caldwell (22) originally differentiated the variety caulivora on the basis of cultural, morphological, and pathogenic differences, as well as the absence of an anamorph. The perithecia of *Diaporthe phaseolorum* var. *caulivora* were described as possessing shorter and more tapering perithecial beaks, which contain smaller asci and ascospores, and are produced on overwintered plant debris in caespitose groups of 2-12 rather than singly as found in southern *Dpc*. Welch and Gilman (541) described the stem canker isolates as being homothallic and void of an asexual pycnidial state. There are several inconsistencies in the literature as to the type of conidia produced when pycnidia are present or if an anamorph associated with the northern stem canker biotype exists at all. Hildebrand (228,229) compared isolates from Indiana and Ontario and found them to be morphologically indistinguishable from one another, except that certain isolates from Ontario produced the imperfect *Phomopsis* stage. Additionally, he reported that perithecia were not always produced in caespitose clusters (as described by Athow) on overwintered debris.

It is clear that genetic diversity reflected by both morphological and physiological differences is common in northern isolates of *D. phaseolorum* var. *caulivora*. In any case, there are reports of different pathogenic biotypes among the isolates collected from cankered southern soybeans. Hobbs (242) indicated that there were two recognizable groups responsible for causing stem canker that could be isolated from soybean. He reported differences in symptom development in southern stem canker and suggested that these differences correlated to cultural differences reported for the two groups. Morgan-Jones and Backman reported that biotypes originating from the southeastern U.S. reach optimal growth at higher temperatures, differ in colony appearance, and produce distinct perithecia and ascospores (356). *Dpc* isolates from the South have been found to characteristically produce pycnidia containing alpha conidia when cultured on certain selective media. However, an associated anamorph has not been reported to occur in nature. Perithecia of southern *Dpc* have been described as being solitary and more or less evenly distributed within the infected lesion of the soybean stem. The perithecial necks have been described as being long (up to 1,500 μm in length), robust, and more or less straight.

Hobbs and Phillips (242) indicated that differences in symptom development and cultural characteristics were significant enough to warrant designation of the disease as "southern stem canker" and isolates of the pathogen as "southern" isolates of *Diaporthe phaseolorum*. Kulik (291) suggested that isolates known to be associated with the stem canker syndrome be referred to as (*forma specialis caulivora*), since cultural and morphological differences were not sufficient enough to warrant varietal separation. Others have suggested that a distinct variety may exist in the southeastern U.S. The current taxonomic status of the causal organism of stem canker in southern soybean is definitely in need of review at this time.

RACES

Along with symptomologic and cultural differences reported between the two geographical biotypes, they also differ in pathogenicity. Keeling (268) reported the presence of several races of Dpc which differed in their pathogenicity on various soybean cultivars. Using the toothpick inoculation method described by Crall (109) on cultivars Kingwa, Tracy-M, Arksoy, Centennial, S-100, and J77-339, he was able to distinguish among six physiological races, three northern and three southern. Higley (226) found that soybean cultivars adapted to Iowa (Blackhawk, Harosoy, and L4404) were susceptible to isolates from that region and resistant to isolates from Mississippi. The cultivar J44-339, which is very susceptible to all southern isolates of stem canker, was resistant to northern isolates of D phaseolorum var. caulivora. Weaver (personal communication) has found that indeterminate soybean cultivars in maturity groups I, II, and III, including those known to be susceptible to northern isolates of Dpc were highly resistant to southern isolates when evaluated using toothpick-inoculation. In comparison, determinate cultivars in maturity groups II, III, and IV in general were found to be susceptible, especially those with the cultivar Ransom in their lineage.

EPIDEMIOLOGY AND CONTROL

Since the first reports of stem canker in the south, the frequency and severity of disease outbreaks has been erratic from one year to the next, even within areas known to be severely infested. Observations made in Alabama (Table 1) serve to illustrate the problem. The development of appropriate control measures for stem canker in the south is dependent on a reliable system for predicting these erratic disease outbreaks. The unpredictable nature of stem canker development has made the understanding of the epidemiology of southern Dpc a major goal of our research program. A diagramatic representation of the stem canker disease cycle follows (Fig. 1).

In our studies, all soybean cultivars were found to be susceptible to infection by Dpc at any developmental stage and time during the growing season (Fig. 2). Further, all plants infected with Dpc remained asymptomatic until they entered the reproductive phase. Additionally, it was determined that plants that were infected during late vegetative or early reproductive stages rarely developed symptoms. However, susceptible plants infected in the early vegetative stages always developed disease.

Table 1. Stem canker disease severity in susceptible cultivars at two Alabama locations between 1980 and 1986.

Year	Location	
	Blackbelt	Tallassee
1980	45%	---
1981	15%	---
1982	30%	<1%
1983	65%	98%
1984	15%	5%
1985	35%	0%
1986	<1%	0%

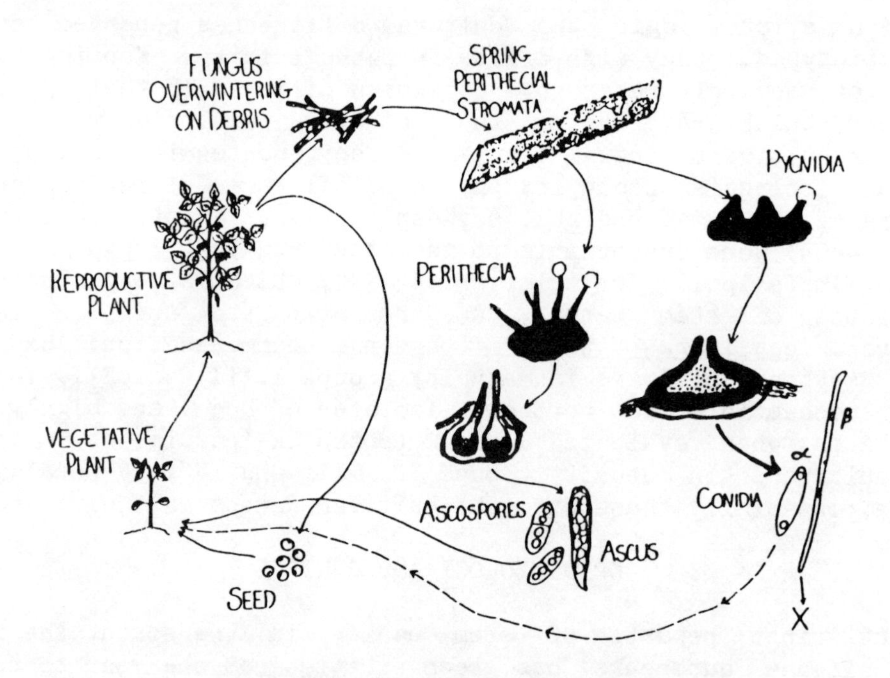

Fig. 1. Disease cycle of Diaporthe phaseolorum var. caulivora on soybean (28).

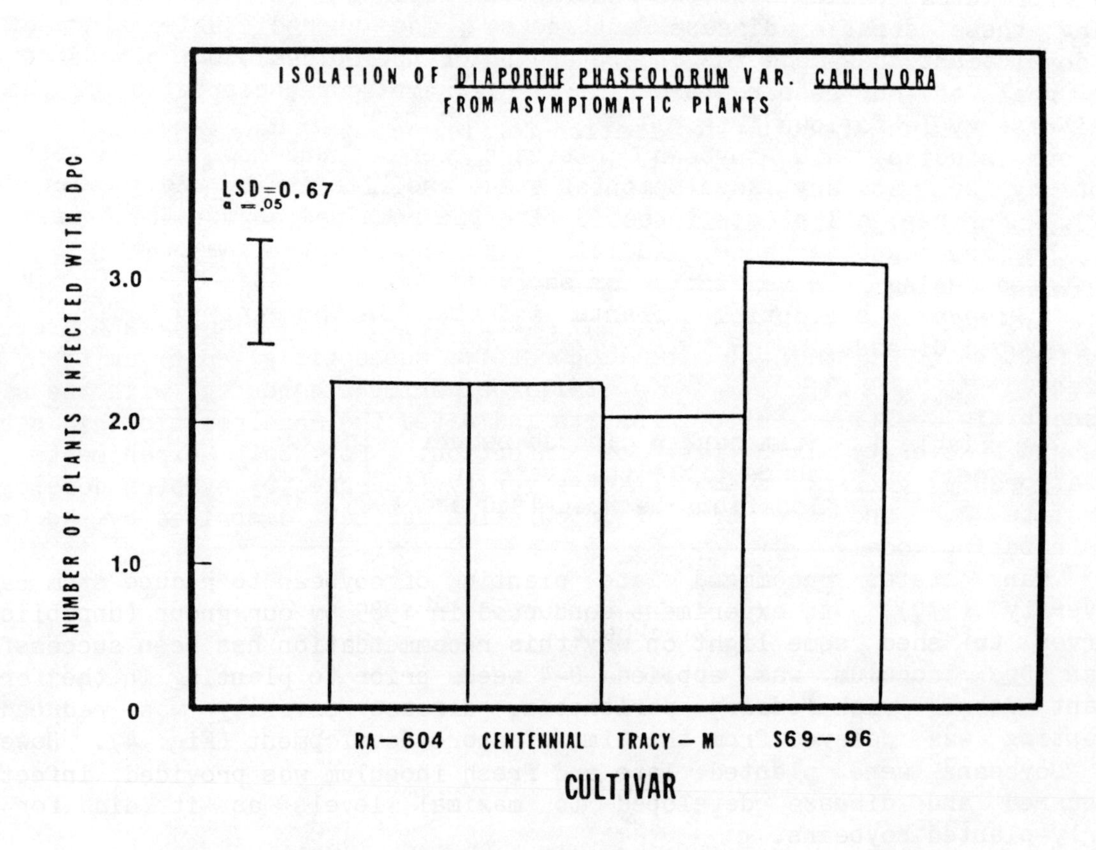

Fig. 2. Soybean plants of four cultivars infected with Dpc at R_3. Numbers reflect means of 20 plants sampled/cultivar/replication (474).

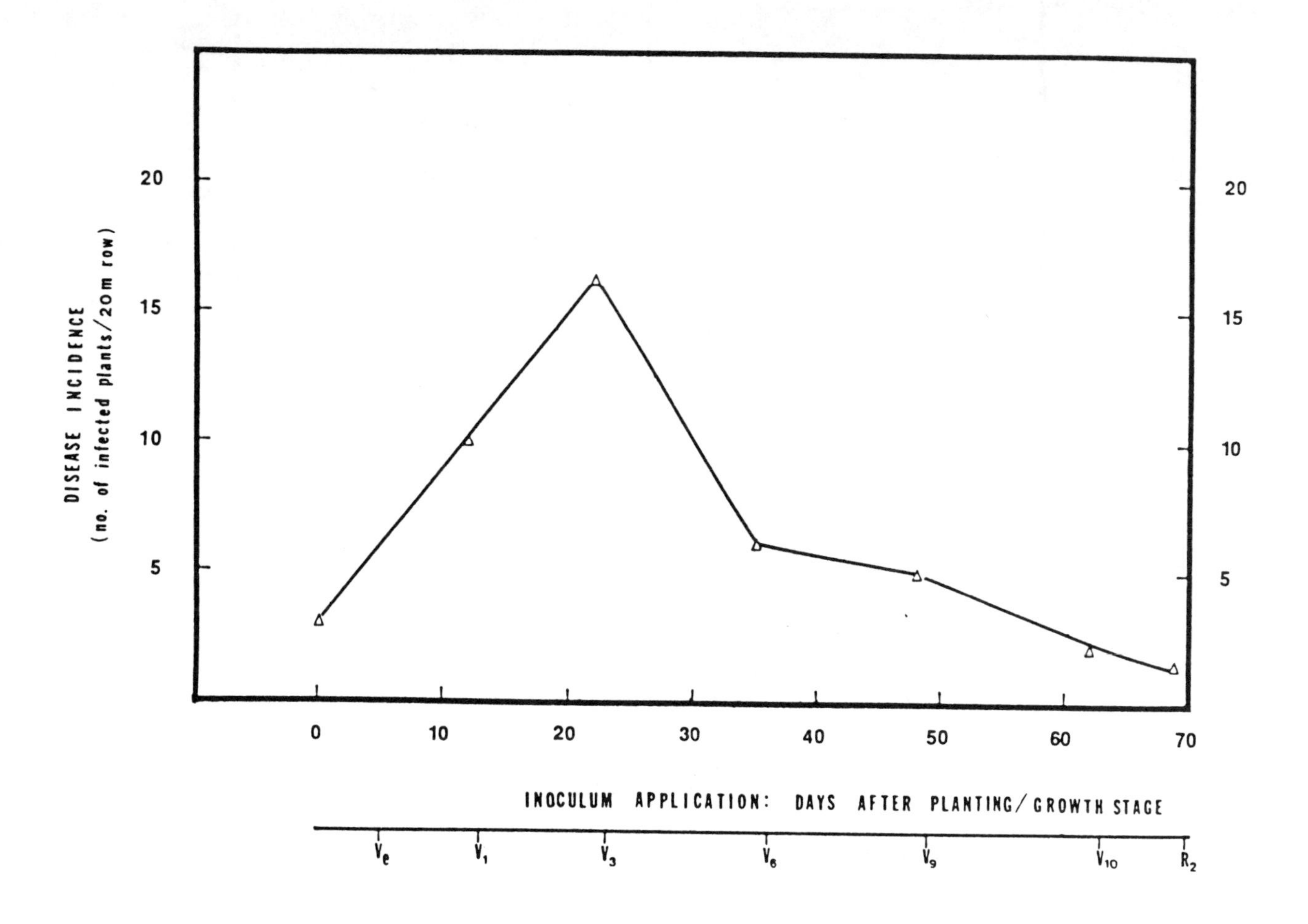

Fig. 3. Time of inoculum application to field-grown Kirby cultivar soybeans in relation to disease severity at R_6 (474).

In one experiment, where Dpc inoculum was applied to soybeans at different growth stages, we related time of infection (relative to plant development) to disease severity (Fig. 3). Maximal disease severity occurred when plants of the moderately susceptible cultivar Kirby were infected at V_3 (152). In a similar experiment conducted with the highly susceptible cultivar Hutton, results indicated the same relationship between disease severity and time of infection. For both experiments, the relationship between stem canker severity (measured by symptom development at late R_6) and the time of inoculation was best described by the Cauchy distribution model.

Many states recommend late planting of soybean to reduce stem canker severity (179). An experiment conducted in 1985 by our group (unpublished) serves to shed some light on why this recommendation has been successful. When Dpc inoculum was applied 0-4 weeks prior to planting in the form of plant debris with oozing perithecia, disease severity was reduced as planting was delayed from the time of spore development (Fig. 4). However, if soybeans were planted late and fresh inoculum was provided, infections occurred and disease developed to maximal levels as it did for the early-planted soybeans.

Retrospective analysis of Alabama weather data since 1977 in comparison to stem canker disease severity in experimental plots has revealed a strong relationship between number of rainfall events and stem canker. The number

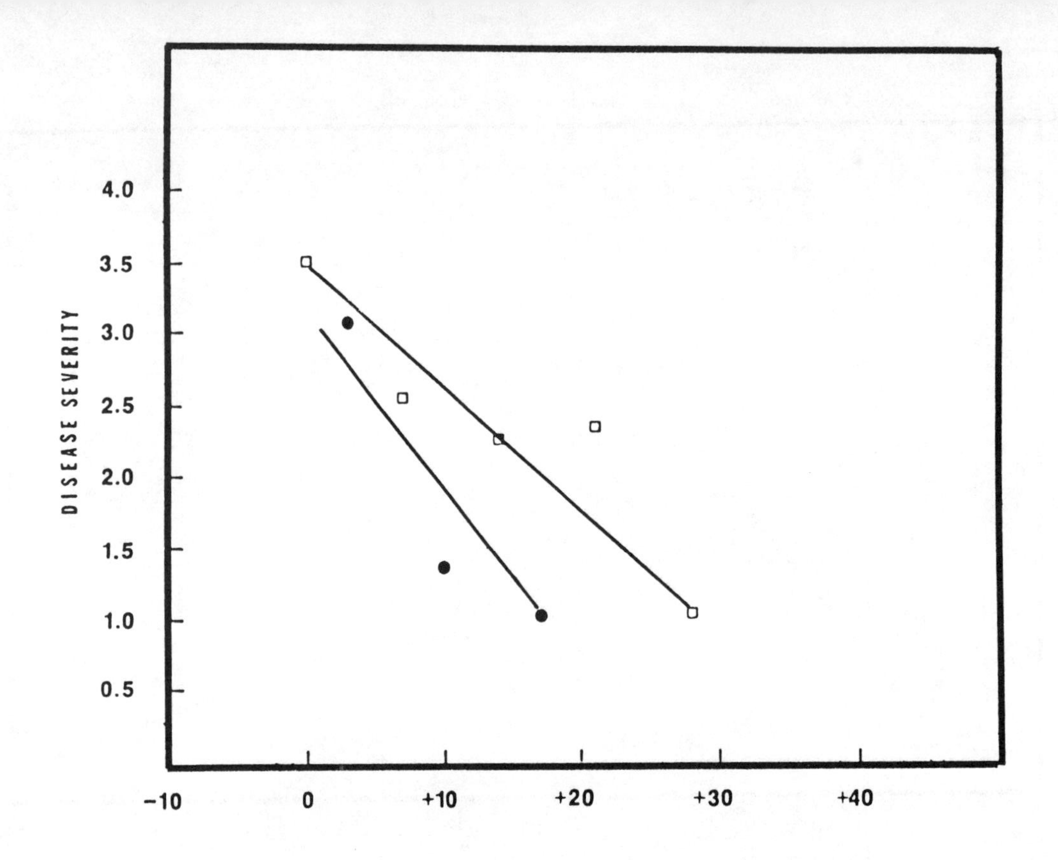

Fig. 4. Time of inoculum application in days prior to planting as it relates to disease severity at R_6 in Hutton cultivar. Disease severity measured by pre-transformed arcsin scale (Table 2) (474).

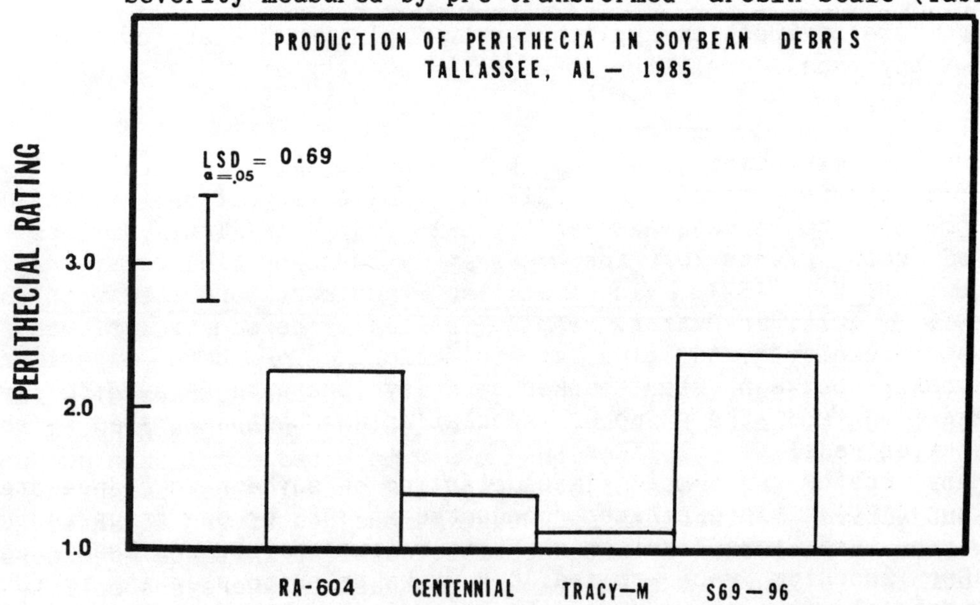

Fig. 5. Development of perithecia on postharvest debris of four cultivars with differing levels of stem canker resistance. Dormant stems rated for degree of perithecial development: 1 = no perithecia, 2 = 25% of stem surface supporting perithecia, 3 = 75% of stem surface supporting perithecia (474).

of rainfall events in 7-day periods extending from 2 weeks prior to planting to bloom was evaluated. Severe disease development was most strongly correlated with high numbers of rainy days in the 7-day period around V_3.

Delayed planting reduces stem canker severity through two mechanisms: 1) perithecia and spores that may have developed in the rains of April and May are mostly exhausted by mid-June; and 2) June is historically one of the driest months of the year in Alabama, and the probability of getting several rainy days in succession is low, giving the soybean plants a good chance of escaping infection until late developmental stages.

Table 2. Pre-transformed arc sine scale for assessing severity of stem canker in soybean fields (28).

Score	Description (per 30-m row)
0.0	No disease
0.2	One plant dead or dying
0.5	Three plants dead or dying
0.7	Seven plants dead or dying
1.0	10% of plants dead or dying
1.3	15% of plants dead or dying
1.5	20% of plants dead or dying
2.0	35% of plants dead or dying
2.4	45% of plants dead or dying
2.5	50% of plants dead or dying
2.7	60% of plants dead or dying
3.0	65% of plants dead or dying
3.4	75% of plants dead or dying
3.8	85% of plants dead or dying
4.0	90% of plants dead or dying
4.5	95% of plants dead or dying
4.8	Two or three live plants
5.0	All plants dead

As was mentioned earlier, all soybean cultivars are successfully infected by Dpc. Resistant cultivars, however, do not allow cankers to grow as large, and they produce significantly fewer perithecia per infected plant than do susceptible cultivars (Fig. 5). The role of overwintering inoculum has not been critically examined. Subjective evaluations of susceptible cultivars planted the year following either a resistant or a susceptible cultivar indicate much more disease where susceptible followed susceptible in contrast to where susceptible followed resistant cultivars. These results indicate that disease severity can be reduced by crop rotation and cultivar rotation. These results are also in keeping with the response of a monocyclic disease to reductions in level of primary inoculum.

The relationship between stem canker severity and yield loss has been well documented (28). The pre-transformed arc sine rating scale (Table 2) relates disease severity at late R_6 to yield loss (Fig. 6).

Control of southern stem canker, like northern stem canker, has centered on eliminating highly susceptible cultivars from production (Table 3). Since there are not enough resistant cultivars available that incorporate other important

Table 3. Relative resistance of selected soybean cultivars to southern isolates of *Diaporthe phaseolorum* var. *caulivora*, causal organism of stem canker (28).

Resistant		Moderately resistant		Moderately susceptible		Susceptible	
Maturity group	Cultivar	Maturity group	Cultivar	Maturity group	Cultivar	Maturity group	Cultivar
I	Blackhawk	V	Terra Vig 505	II	Gnome	V	AP 55
	Hardin		Deltapine 105	III	Elf		A 5539
	Hodgson 78		Deltapine 345		Sprite		RA 502
II	Century		Wilstar 550	IV	Pixie	VI	RA 604
	Corsoy 79		Shiloh	V	Bedford		Brysoy 9
	Hawkeye	VI	Centennial		Forrest		Bradley
III	Cumberland		Davis		Essex	VII	Coker 237
	Williams 82		RA 680		A 5474		Bragg
V	Bay		Coker 156	VI	Jeff		McNair 770
VI	Tracy-M	VII	Wright		Lee 74		Wilstar 790
VII	Braxton		Ransom		S69-96		RA 701
VIII	Dowling		Coker 317		Deltapine 506	VIII	Coker 338
			GaSoy 17	VII	Gregg		Hutton
		VIII	Coker 368		Gordon		RA 801
			Coker 488	VIII	Foster	IX	Jupiter R
			Cobb		Kirby		Santa Rosa R

[a]Reactions of cultivars in maturity groups I, II, III, IV, and IX are based on greenhouse inoculation tests; all others are based on field tests.

agronomic characteristics, southern farmers must learn to manage the disease based on an understanding of disease epidemiology. When cultivars of intermediate susceptibility are to be planted, they should be planted on rotated fields, or fields planted to a resistant cultivar the previous year. As a second precaution, farmers should plant 3 weeks after perithecia have been known to mature. Our observations in Alabama have shown that over the last three years spore release has occurred after mid July, too late to result in disease development. Needless to say, a farmer cannot delay planting to such dates. Should the farmer plant at a more appropriate time, followed by a prolonged rainy period that leads to Dpc infection in the early vegetative stages, then he may need to use fungicides to reduce disease severity. To date, the best fungicides have been found to be benzimidazoles (28); the sterol inhibitor class has been uniformly poor in controlling stem canker. Single applications of benzimidazoles, banded over the small plants, have greatly reduced disease when applied shortly after infection.

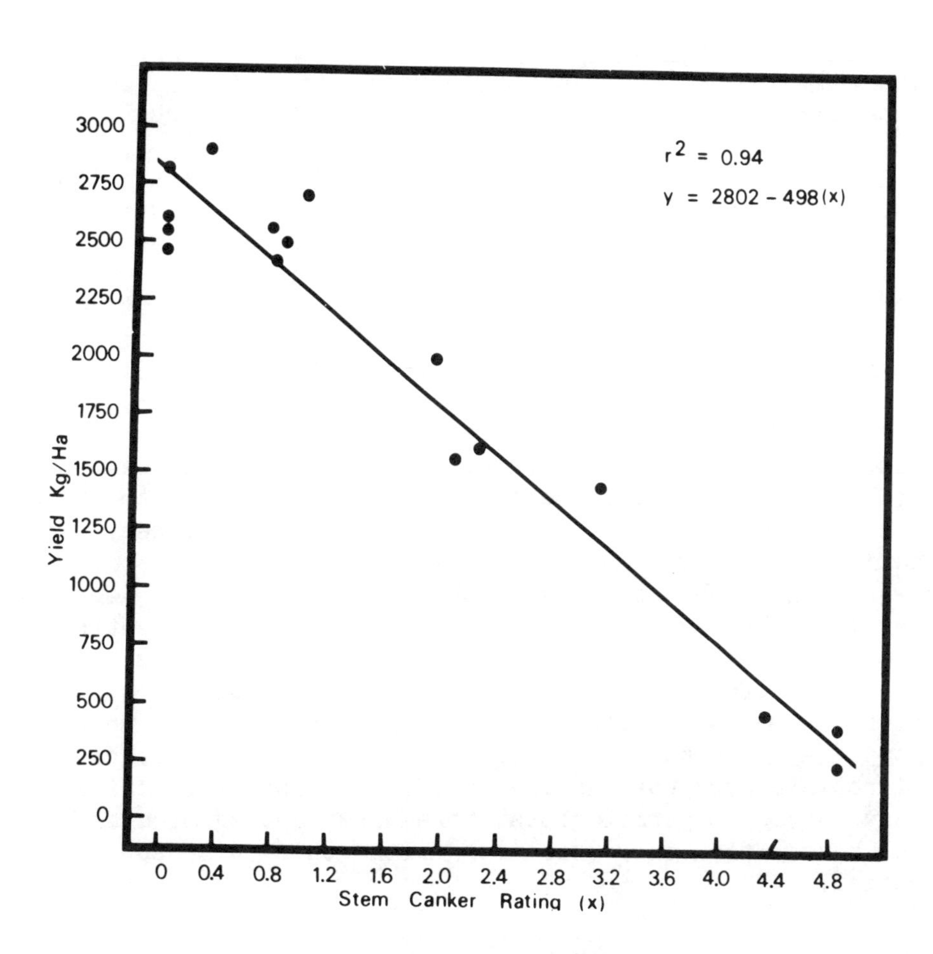

Fig. 6. Relationship between soybean yield and stem canker severity according to the pre-transformed arc sine rating system (Table 2) (28).

SCLEROTINIA STEM ROT OF SOYBEAN

Craig R. Grau

Department of Plant Pathology
University of Wisconsin-Madison
Madison, WI 53706

Sclerotinia stem rot of soybean occurs throughout the world but appears to have its greatest impact on production in the temperate regions of North America and South America (469,515,565). In North America, Weiss (541) first reported the disease to be present in Iowa, New York, Maryland and Virginia. Since 1946, Sclerotinia stem rot has been reported in Arizona (237), Illinois (88,108), Minnesota (189), New Hampshire (88), Ontario (227), Wisconsin (190), Virginia (395), and personal communications indicate the disease is present in Indiana and Ohio. Sclerotinia stem rot has not been reported in the typically hot soybean growing areas of the southern United States. Nationally, the disease is considered to be minor because it does not involve a high percentage of the total soybean acreage. However, Sclerotinia stem rot can result in extensive plant mortality during pod development resulting in significant yield reduction.

Chamberlain (88) was the first to make a detailed report on Sclerotinia stem rot in the Midwest after he observed localized, but severe, outbreaks of the disease in Illinois in 1946. Chamberlain (88) summarized his findings by the following quote; "There appears to be no ready explanation as to why Sclerotinia stem rot, certainly one of the least prevalent of soybean diseases, can cause such severe but localized damage." After 40 yr, more is known about factors that impact on the incidence and severity of this disease, but an element of mystery still remains as to why sudden outbreaks occur. Cultural practices associated with soybean production have changed in recent years and such changes may be associated with the increased recognition of Sclerotinia stem rot. In addition, soybean production has expanded into areas of the upper Midwest where other hosts of S. sclerotiorum are frequently grown in rotation with soybean. Although the interest in Sclerotinia stem rot has increased, the disease is still treated as a curiosity by many soybean pathologists, breeders, and agronomists. However, a person needs to have only one encounter with this disease to realize its destructive potential.

CAUSAL ORGANISM

Sclerotinia sclerotiorum (Lib.) d By. (=*Whetzelinia sclerotiorum* (Lib.) Korf and Dumont (469) is the primary cause of Sclerotinia stem rot of soybean. *S. sclerotiorum* (285) is characterized by fluffy white mycelium and black sclerotia that are formed by aggregations of mycelium. Sclerotia usually are 2-20 mm long (285), form on and within diseased tissue, and function as resting structures that enable this fungus to survive in soil or mixed with seed. Sclerotia germinate at, or slightly below the soil surface by producing apothecia which in turn produce ascospores which are forcibly ejected and wind disseminated. Steadman's (481) review of white mold of bean provides an illustrated description of the structures and life cycle of this pathogen.

It is important to note that Phipps and Porter (395) found *S. minor*, a species related to *S. sclerotiorum* and a major pathogen of peanut, to cause a similar disease on soybean when grown in rotation with peanut in Virginia. *S. sclerotiorum* also was found, but *S. minor* was more common, and the two species were not found to cause disease in the same field.

SYMPTOMS

Symptoms of Sclerotinia stem rot typically appear during the early stages of pod development (growth stages R3-R4). At the canopy level, foliar symptoms are the first indication that the disease is present. Foliar symptoms are chlorosis and wilt, with tissues between major leaf veins developing a gray-green cast while vein tissues remain green. In time, leaves become totally necrotic, tattered and curled, but remain attached to the stems past maturity. Foliar symptoms of Sclerotinia stem rot could be mistaken for late season Phytophthora root rot, brown stem rot, and stem canker, but differences in stem symptoms among these diseases can be used for diagnosis.

Initially, stem lesions develop at nodes and appear gray and water-soaked. The pathogen rapidly progresses into stem tissues above and below nodes and causes lesions that are 6-30 cm in length and usually encompass the entire stem (190,469). White fluffy mycelium covers the lesion area, especially during periods of high relative humidity. The characteristic black sclerotia of *S. sclerotiorum* are differentiated from mycelium and in time are readily observed on the lesion surface. Initially, lesions are tan and progressively become white, and present a sharp contrast at the interface with green stem tissues. By crop maturity, stem tissues are white and tissues have a shredded appearance if disturbed, and a reddish discoloration is frequently interspersed within diseased stem tissues and at the border of lesions. At harvest, diseased stems are characterized by poor pod development, a white appearance, and an abundance of sclerotia within the pith. Diseased pods are outwardly white in appearance, mycelium and sclerotia are readily observed inside, and infected seed will appear white and moldy. Sclerotia are commonly observed with the harvested grain, and, if free water is present, can cause seed decay problems in storage.

ECOLOGY AND EPIDEMIOLOGY

Disease Cycle

Data on the epidemiology of Sclerotinia stem rot of soybean are very limited. However, data on the disease cycle and epidemiology of white mold of green and dry beans appear to be applicable for Sclerotinia stem rot of soybean (1,481). *S. sclerotiorum* survives as sclerotia that must be at or within 5 cm of the soil surface to produce apothecia (469,481). Apothecia optimally form at soil temperatures of 15-18 C and a water matrix potential of -0.25 bars for 10-14 days (481,509). Apothecia produce ascospores that are forcibly ejected and disseminated by air currents to the host surface. Canopy temperatures less than 30 C and plant surface wetness for 12-16 hr recurring on a daily basis or continuous surface wetness for 42-72 hr are environmental conditions needed for disease development (481). In Wisconsin, mean min/max temperatures of 20/34 C 2 wk before and 2 wk after flowering resulted in the complete absence of Sclerotinia stem rot. The disease was prevalent the previous and following year at the same location when mean min/max temperatures were 12/22 and 18/30 C, respectively (191). Irrigation was employed each year of the study; thus, moisture was unlikely a limiting factor. Although soil and canopy moisture are regarded as most critical, temperature appears to have a dramatic effect on the development of Sclerotinia stem rot.

Blossoms are a primary site of infection by S. sclerotiorum for green and dry beans (2,481,488). Soybean blossoms also are reported to be colonized and serve as a nutrient base for infection of soybean stems when artificially inoculated (100,488), but this situation has not been reported for naturally infected plants. In Wisconsin, S. sclerotiorum first was isolated from blossoms that adhered to the tips of newly emerged pods. Experiments were conducted to follow the progression of plant parts colonized by S. sclerotiorum by selecting one symptomless plant from each of 20 sampling sites at four sampling dates in a plot naturally infested with the pathogen. Results of this study indicate that pods were colonized before nodes and stems as asymptomatic plants (Fig. 1). Recovery from pods increased with time as did recovery from nodes and stems. These data suggest that ascospores germinate and colonize blossom tissues which in turn provide a food base needed by the pathogen to invade pod tissues and progress into the nodal, and eventually stem, tissues. Infection typically occurs at nodes located 10-24 cm above the soil line. Infection of tissues other than nodes such as leaves, petioles, and internodes can occur when plants come in contact with diseased adjacent plants. This is more common when the crop is lodged. The pathogen causes lesions that completely encircle the stem and disrupt the transport of water, mineral nutrients, and photosynthates to developing pods. Sclerotia are formed as host tissues are depleted of nutrients needed by the pathogen. Sclerotia are returned to the soil at harvest or as contaminants associated with seed at planting. The importance of infected seed in the epidemiology of this disease has not been studied, but is likely to be of less importance than sclerotia mixed with seed.

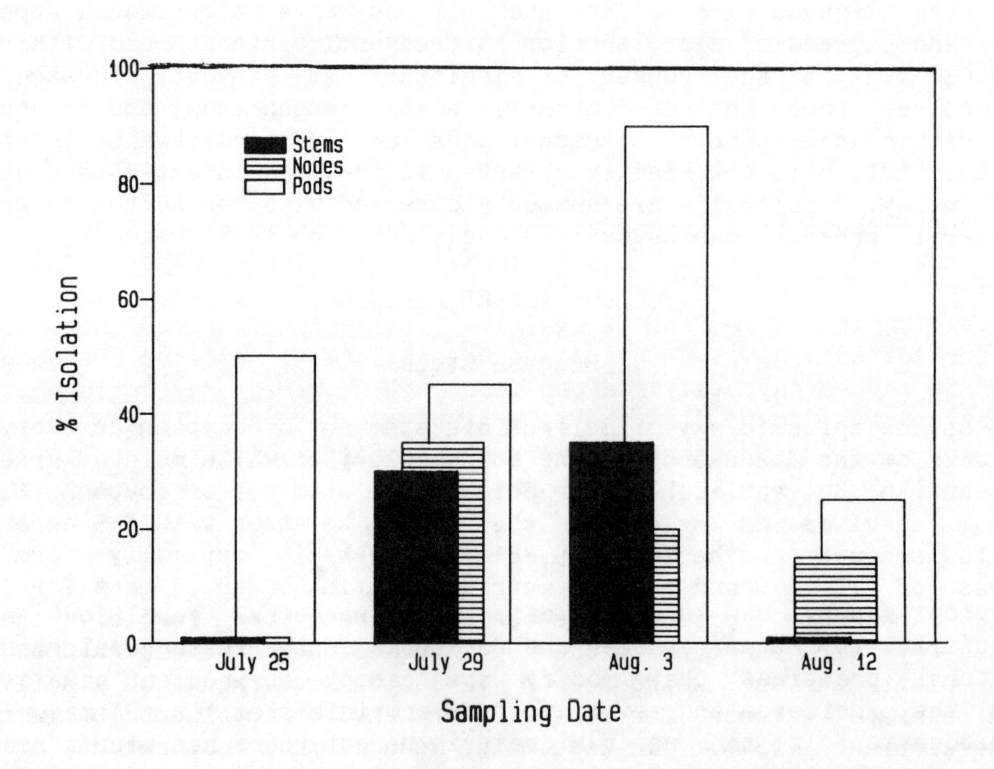

Fig. 1. Percentage of plants from which Sclerotinia sclerotiorum was isolated from pods, nodes or stems at four sampling dates beginning at early pod development.

Hosts of *Sclerotinia sclerotiorum*

S. sclerotiorum is reported to infect almost 400 species of plants (404). Sclerotinia stem rot is relatively uncommon in a corn-soybean rotational scheme. However, the probable occurrence of Sclerotinia stem rot is greatly enhanced by the inclusion of another host crop in the rotation. Hosts are presented in Table 1 that could potentially be grown in rotation with soybean. Sudden outbreaks of Sclerotinia stem rot have been observed in fields monocultured to corn for 5-11 years prior to soybean being planted (C. R. Grau, personal observations). Broadleaf weeds such as lambsquarter (*Chenopodium album* L.), red-rooted pigweed (*Amaranthus retroflexus* L.), velvet leaf (*Abutilon theophrasti* Medic.), and common ragweed (*Ambrosia artemisiifolia* L.) were observed to be diseased in these fields and could have served as a source of inoculum as suggested by studies in Brazil (565). Corn, small grains, and grass weeds are reported as hosts of *S. sclerotiorum*, but personal experience indicates this is rare.

Soybean Cultivar

Soybean cultivars differ in reaction to Sclerotinia stem rot (53,100,189,190,191). Cultivars range from moderately resistant to very susceptible within maturity groups 0 through II (Tables 2, 3). However, less resistance is reported within maturity group III and later (100). Some question still remains whether physiologic resistance functions against *S. sclerotiorum* or differences in cultivar reactions are due to other factors such as plant architecture, maturity, and lodging characteristics. Some of the least-diseased cultivars are within maturity group 0, which may suggest that these cultivars escape infection because of their generally smaller stature or tendency to flower earlier. However, very susceptible cultivars can also be found in this group (Table 2). Cultivars that characteristically lodge tend to express greater disease, but lodging is not totally related to a cultivar's reaction to Sclerotinia stem rot. For example, Corsoy 79 readily lodges, but usually is much less diseased than Wells II, a less-lodging cultivar. The expression of apparent resistance of soybean to *S. sclerotiorum* is greatly affected by cultural practices such as row width (191). This latter situation causes difficulties in standardizing the reactions of cultivars to *S. sclerotiorum*. Whether or not a physiological basis for resistance against *S. sclerotiorum* functions in soybean is a major unresolved issue. Sutton and Deverall (489) reported the occurrence of glyceollin in soybean hypocotyls after inoculation with *S. sclerotiorum*. However, no information was found in the literature on whether cultivars that differ in reaction to *S. sclerotiorum* have corresponding differences in glyceollin concentration that relates to the host phenotype.

Crop Canopy Modification

The crop canopy can greatly affect environmental conditions needed for optimum activity by the pathogen and for subsequent disease development (491). Thus, cultural practices that modify the canopy environment have a potential impact on the incidence and severity of Sclerotinia stem rot. A major change in soybean production in the upper Midwest in recent years has been a reduction of row widths from 75-90 cm (conventional) to 18-38 cm (narrow). Yields from narrow row plantings are often 20% greater compared to yield from conventional row width systems (108,374). However, a 65% increase in incidence of Sclerotinia stem rot and a 42% reduction in yield has been measured for soybean cultivars grown in narrow, as compared to wide row widths (191). Plant population as well as row width also affects the incidence of Sclerotinia stem rot (565). Sprinkler irrigation is another cultural practice that can greatly modify the canopy

Table 1. A partial list[a] of agronomic and vegetable crops reported to be hosts of *Sclerotinia sclerotiorum*.

Crop	Scientific Name
Table Beet	*Beta vulgaris* L.
Rapeseed (Canola)	*Brassica napus* L.
Cole Crops (Cauliflower, etc.)	*Brassica oleracea* L.
Watermelon	*Citrullus vulgaris* Schrad.
Muskmelon	*C. melo* L.
Cucumber	*C. sativus* L.
Winter Squash	*C. maxima* Dcne.
Pumpkin	*C. pepo* L.
Summer Squash	*C. pepo* var. *melopepo* (L.) Alef.
Peppermint	*Mentha piperita* L.
Crownvetch	*Coronilla varia* L.
Soybean	*Glycine max* (L.) Merr.
Lentil	*Lens culinaris* Medik.
Birdsfoot Trefoil	*Lotus corniculatus* L.
Alfalfa	*Medicago sativa* L.
Sweetclovers	*Melilotus* spp.
Sainfoin	*Onobrychis viciifolia* Scop.
Scarlet Runner Bean	*Phaseolus coccineus* L.
Lima Bean	*P. limensis* Macf.
Green and Dry Bean	*P. vulgaris* L.
Pea	*Pisum sativum* L.
Field Pea	*P. sativum* L. subsp. *arvense* Poir
Clovers (Red, White, etc.)	*Trifolium* spp.
Cowpea	*Vigna sinensis* (Torner) Savi
Onion	*Allium cepa* L.
Flax	*Linum flavum* L.
Cotton	*Gossypium hirsutum* L.
Peanut	*Arachis hypogaea* L.
Buckwheat	*Fagopyrum esculenthum* Moench
Tomato	*Lycopersicum esculentum* Mill.
Tobacco	*Nicotiana tabacum* L.
Potato	*Solanum tuberosum* L.
Carrot	*Daucus carota* L. var *sativa* DC.
Parsnip	*Pastinaca sativa* L.
Sunflower	*Helianthus annuus* L.
Lettuce	*Lactuca sativa* L.

[a]Taken from a host range list prepared by H. F. Schwartz, Dept. of Botany and Plant Pathology and Weed Science, Colorado State University, Fort Collins, CO 80523.

Table 2. Disease severity indices (DSI)[a] for soybean cultivars evaluated for resistance to <u>Sclerotinia sclerotiorum</u> in naturally infested plots in Wisconsin.

Cultivar	Maturity Group	DSI 1981		DSI 1982		Mean
		Loc. 1	Loc. 2	Loc. 1	Loc. 2	
McCall	0	4	2	5	0	3
Evans	0	52	36	53	23	41
Hodgson 78	I	27	11	40	31	27
Hardin	I	53	29	--	34	39
Norman	0	3	1	--	--	2
Maple Arrow	0	3	0	--	--	2
Maple Presto	0	5	0	--	--	3
Altona	0	13	1	--	--	7
Wilken	0	21	9	--	--	15
Swift	I	57	54	--	--	56
Lakota	I	59	20	--	--	40
Ozzie	0	--	--	13	13	13
Dawson	0	--	--	86	31	59
Simpson	0	--	-	60	57	59
LSD P = 0.05		18	15	16	12	

[a]Disease severity index based on a scale of 0-100; 0 = no symptoms present to 100 = all plants have extensive stem lesions and pod development is severely reduced (190).

Table 3. Summary of disease reactions for soybean cultivars evaluated for resistance to Sclerotinia stem rot at 38 cm row widths in Wisconsin.

Cultivar	Maturity Group	DSI[a]	No. trials	Cultivar	Maturity Group	DSI[a]	No. trials
Corsoy	II	15	2	Amcor	II	42	2
Hodgson	I	21	2	Century	II	46	2
Hodgson 78	I	27	6	Amsoy 71	II	43	2
Hark	I	33	1	Wells	II	54	2
Sprite	II	35	1	Wells II	II	57	1
Vickery	II	37	2	Wayne	III	62	2
Evans	0	39	4	Beeson 80	II	68	1
Corsoy 79	II	40	1	Beeson	II	74	1
Hardin	I	40	4	Gnome	II	76	2

[a]Disease severity index of 0-100 (191).

environment and lead to greater disease, especially if applied when flowering is occurring at the lower nodes (191).

Effect of Tillage

Williams and Stelfox (552) reported that deep plowing reduced the number of apothecia that formed under a crop canopy when compared to shallow tillage in one of two years. However, a 3 year study in Brazil showed a consistent increase in disease incidence for soybean grown in soils that were not tilled compared to deep plowing (565). The numbers of apothecia produced per unit area of soil surface was not measured in the latter study. The survival and activity of sclerotia is greatly dependent upon soil moisture, and, more importantly, the range of soil moisture extremes (481). Thus, rainfall and the density of the crop canopy could strongly interact with the effects of tillage on the activity of S. sclerotiorum.

Effect of Herbicides

Soil-applied herbicides have been shown to affect mycelial growth (84) and carpogenic germination of sclerotia of S. sclerotiorum (78,406). Herbicides commonly used for soybean production, such as trifluralin, pendimethalin, and metribuzin, stimulate carpogenic germination of this soil-borne plant pathogen (406). However, the effect of this stimulation on carpogenic germination has not been studied in relation to its effect on disease severity. Atrazine and simazine, herbicides commonly associated with corn production, stimulate sclerotia to germinate, but apothecia develop abnormally and do not produce asci and ascospores (78,406). Thus, corn culture may reduce soil inoculum of S. sclerotiorum by several mechanisms: corn not being a host; associated herbicides provide broadleaf weed control which reduces the population of potential hosts; and herbicides stimulate sclerotia to germinate, but abnormally resulting in reduced reproduction which in turn leads to a depletion of sclerotia in the soil.

Biotic Soil Factors

Several soilborne fungi have been shown to invade sclerotia of S. sclerotiorum, Sporidesmium sclerotivorum and Coniothyrium minitans are both reported as parasites of sclerotia and can affect the length of time that sclerotia survive in the soil (24,248). Species within the genera Trichoderma and Gliocladium also are active as invaders of sclerotia, with G. virens most frequently associated with sclerotia and reduced carpogenic germination of S. sclerotiorum (358).

Impact on Yield

No extensive disease loss assessment studies have been conducted for Sclerotinia stem rot of soybean. Disease incidence and severity have been used to measure Sclerotinia stem rot (189,190,191), but, generally, disease severity differs very little among infected plants for a specific cultivar. Thus, disease incidence and disease severity, based on individual plant assessment, are closely related. General field observations would indicate that considerable yield reduction would occur because of the destructive effect of the disease on an individual plant basis and the timing of symptom expression. Seed production by infected plants can be completely inhibited, but generally some seed are set although they are usually reduced in size. Also, diseased plants are killed at a stage of crop development (R5-R6) when uninfected plants do not readily compensate for lost yield from diseased plants. Data from five epidemics in Wisconsin indicates yield was reduced 0.066-0.398 Mg/ha (mega grams)/ha if 10% plant mortality is caused by Sclerotinia stem rot (Fig. 2). The mean for the five epidemics was 0.258 Mg/ha for each 10% increment of disease incidence (Table 4). Disease incidence was determined by the percentage of plants that expressed stem lesions caused by Sclerotinia stem rot. Disease severity has been determined by rating the stem lesions on a 0-3 scale: 0 = no symptoms, 1 = lesions only on lateral branches, 2 = lesions on main stem, but pod development was not significantly reduced, and 3 = lesions on main stem and pod development was poor. This data was converted to a 0-100 index (191). The use of a disease severity rating improves the regression equation slightly (Fig. 2) when compared to the use of disease incidence (epidemic no. 1 in Table 2). However, the use of a disease severity rating is more time consuming and may not be practical in some situations.

INTEGRATED CONTROL

Many of the factors discussed in the previous section can be modified to reduce the risk of yield loss due to Sclerotinia stem rot. The specific combination of management practices implemented would depend greatly on specific situations.

Sclerotinia stem rot is more prevalent when grown in rotation with other hosts and when management practices are designed to achieve maximum yields. For example, planting at reduced row widths, irrigation, and high soil fertility all are implemented because of greater yield potentials, but all can result in a dense canopy that also favors the development of Sclerotinia stem rot. Each management practice presents a potential risk, but specific modifications can usually reduce the overall risk factor.

Because soybean cultivars differ in susceptibility to Sclerotinia stem rot, cultivar selection can greatly reduce the risks associated with specific management systems. For example, highly susceptible cultivars should never be planted in reduced row widths or be planted immediately after another host crop in a field infested with _S. sclerotiorum._ Most cultivars released by commercial companies have not been evaluated for resistance to Sclerotinia stem rot. Resistance has been reported for public cultivars within maturity groups 0-III (100,189,190,191 and Tables 3 & 4), with the greatest level of resistance being associated with group 0 cultivars. The disease reaction of a specific cultivar can be influenced by climatic conditions and management practices (191). For high disease potential situations, narrow row culture should be discontinued and a cultivar with a low disease reaction should be selected for planting in order to achieve the lowest risk situation. In many cases, cultivars with low disease reactions are not available or are undesirable because of poor agronomic traits. In this situation, just the shift from narrow to wide culture may be enough to lower the risk of yield loss due to Sclerotinia stem rot. Many growers may not be aware that they have the potential for this disease until they change a cultural practice, such as row width.

Benomyl and thiophanate (methyl) are registered as foliar fungicides for soybeans and are effective against _S. sclerotiorum_ on similar crops such as snap beans. Studies conducted in Wisconsin indicate that one application of benomyl (Benlate 50W at 1 kg/ha) can reduce Sclerotinia stem rot (C. R. Grau, unpublished data). However, fungicides must be applied when soybeans are producing flowers or as pods are just emerging on the lower one-half of the plant. Thus, timing and penetration of the fungicide through the soybean canopy present problems in their effective use for control of Sclerotinia stem rot. Currently the label does not permit this early application.

A summary of integrated control tactics are presented in Table 5.

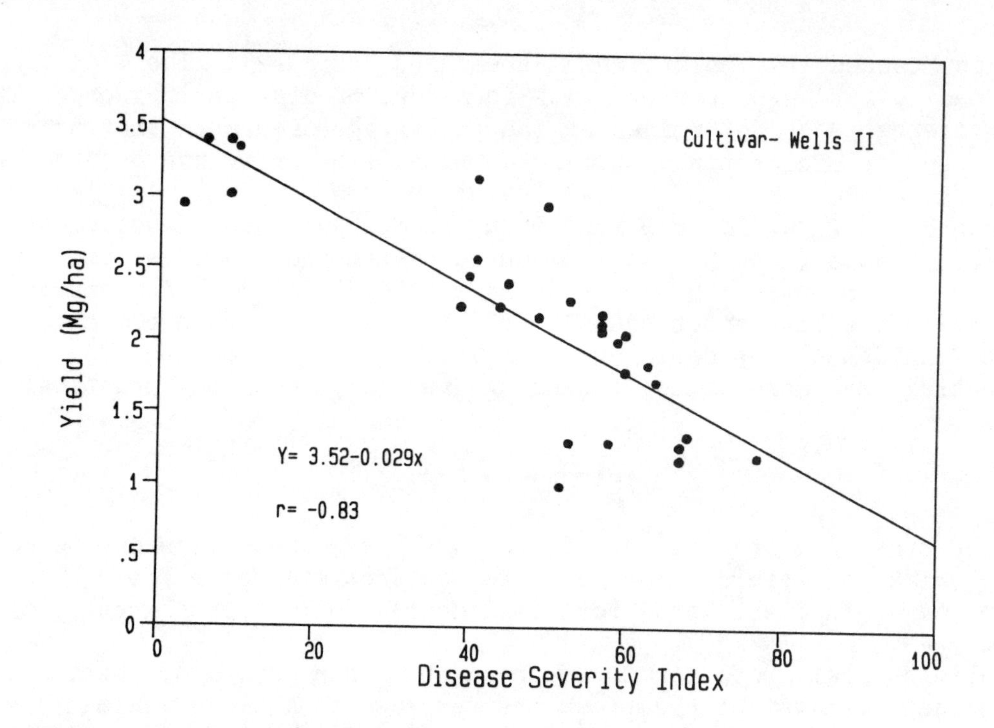

Fig. 2. Relationship between disease severity of Sclerotinia stem rot and yield of soybean.

Table 4. Relationship between percent incidence of Sclerotinia stem rot and soybean yield in five epidemics.

Epidemic[a]	$\bar{x}$ Disease[b] incidence (%)	Yield kg/plot	Regression equation	t-ratio[c]	R^2
1	49	2.2	3.39-0.0237x	5.71*	52.1
2	49	2.7	4.60-0.0398x	3.15*	32.0
3	27	3.6	4.23-0.0236x	7.51*	70.6
4	41	3.0	4.65-0.0353x	9.11*	63.5
5	39	2.8	3.43-0.0066x	2.38*	53.8

[a]Epidemic no. 1 employed the cultivar Wells II and disease incidence was measured in 30 plots. Epidemics no. 2-3 employed the cultivar Wells II and disease incidence and yield were measured in 36 and 54 plots, respectively. Epidemic no. 4 employed the cutivars SRF-200, Asgrow 2656, Wells, Corsoy, Hodgson, and Steele and disease incidence and yield were measured in 24 plots. Epidemic no. 5 employed the cultivar Evans and disease incidence and yield were measured in 20 plots. Plots were 1.85 x 6.2 m and consisted of four rows spaced 38 cm apart. Disease incidence and yield data from each plot was regressed against each other.

[b]Disease incidence was determined by the percentage of plants that expressed stem lesions at growth stage R7 (full pod development).

[c]An asterisks indicates statistical significance at $\underline{P}$ = 0.10.

Table 5. A summary of tactics available for integrated control of Sclerotinia stem rot of soybean.

Tactic	Comments
Cultivar Selection	- Soybean cultivars range from moderately resistant to very susceptible within maturity groups 0 through III. - Disease reactions of specific cultivars differ with changes in cultural practices and climatic conditions. - Cultivars that characteristically lodge more tend to express greater disease, but lodging is not totally related to a cultivar's reaction to Sclerotinia stem rot. - Physiologic races have not been reported for _S. sclerotiorum_.
Water Management	- Greater disease is associated wih soils with high water holding capacity. - Severe disease can develop in sandy soils if sprinkler irrigation is applied.
Canopy Modification	- Avoid row widths less than 76 cm if Sclerotinia stem rot potentials exist. - Cease irrigation during the flowering period. - Avoid cultural practices that promote lodging.
Crop Sequence	- In most cases, Sclerotinia stem rot becomes a problem after moderate to high levels of disease in another host crop grown in rotation with soybean. Plant a non-host 1-2 years between soybeans and other host crops.
Crop Rotation	- Crop rotation can supplement other control practices, but alone is not an effective control.
Field Monitoring	- Early detection of Sclerotinia stem rot in soybeans or other crops in rotation with soybean allows time to adjust management practices to combat the disease. Sclerotinia will usually be detected in other crops before soybean.
Weed Control	- Control broadleaf weeds in all crops, because many are hosts of _S. sclerotiorum_. In addition, weeds can create a more dense canopy which in turn favors disease development.
Plant Nutrition	- Plant nutrition has not been shown to have an direct affect on Sclerotinia stem rot, but fertilization practices may promote lodging.
Fungicides	- No fungicides are specifically registered for control of Sclerotinia stem rot of soybean. - Benomyl and thiophanate (methyl) are registered for control of several soybean diseases and do provide for control of Sclerotinia stem rot if applied at early flower.

FUTURE RESEARCH NEEDS

The development of resistant cultivars within northern maturity groups is a critical need for the soybean industry. This objective is greatly curtailed by the lack of understanding of how resistance functions and is inherited. Actually, there is debate over whether physiological resistance to S. sclerotiorum actually exists within soybean germplasm. I believe resistance does function, but many breeders and pathologists maintain that measured differences between cultivars for disease incidence and severity is a function of plant architecture, flowering date, or other factors that allow the plant to escape infection. Research is needed to resolve this issue, because the factor(s) responsible for host responses would influence breeding and selection techniques.

Soybean cultivars can be evaluated for their reaction to Sclerotinia stem rot in the field (189,190,191) or in controlled environments (53,100). Currently, techniques to evaluate soybean lines for resistance are relatively consistent but are often time and space consuming, and refinement of these techniques would be of value. Consistent disease reactions of cultivars from test to test appears to be a problem with most techniques. I speculate that one source of the problem is that many soybean cultivars and lines are heterogenous for resistance to this pathogen. Data presented in Fig. 1 shows that 90% of the plants sampled had infected pods at the third sampling date. However, a final disease incidence of 49% was measured at physiological maturity. Although the pods were colonized, theoretically the fungus did not cause stem lesions in 31% of these plants. Small restricted lesions (0.5-1 cm) are commonly observed at nodes, especially for cultivars with lower disease reactions. Lesions are usually reddish-purple, indicative of anthocyanin accumulation often associated with resistant responses. In this situation, it is believed that a pod was colonized allowing the pathogen to progress to the node. However, an apparent physiological response by the host restricted the colonization of the stem. Such plants are observed in many cultivars and it might be possible to select within cultivars for the improvement of Sclerotinia stem rot resistance. The issue of heterogeneity is important when selecting plants for crosses in breeding programs and studies on the inheritance of resistance. Potential parent plants should be evaluated for their reaction to S. sclerotiorum before their use in soybean cultivar development.

Sclerotia of S. sclerotiorum are commonly associated with seed harvested from diseased plants. The role of contaminated seed lots and infected seed has not been studied for the introduction of the pathogen into uninfested fields and the general epidemiology of the disease. Sclerotia as contaminants are likely to be more of a factor than are infected seed. Infected seed are usually reduced in size and appear to be lost in the process of mechanical harvesting and seed conditioning. However, research is needed to address this potential problem.

SOYBEAN SUDDEN DEATH SYNDROME

Donald H. Scott

Department of Botany and Plant Pathology
Purdue University

Sudden death syndrome (SDS) is a relatively new disorder of soybeans. To date, the causal agent(s) of SDS have not been determined. The name is descriptive in that normal-appearing plants turn yellow and die rather quickly, in somewhat circular to elongated patches of a field, after pod set.

SDS was first observed in Arkansas in 1971, but caused little concern until 1982 when it caused an estimated 25% yield reduction in 5 to 10% of Arkansas' soybean acreage (Hirrel, M. C., and Rupe, J., Personal Comm.). Since 1982, SDS has been identified in Illinois, Indiana, Kentucky, Mississippi, Missouri and Tennessee (221,240,453,564, and Hirrel, M. C., and Rupe, J., Personal Comm.). SDS was first identified in Indiana in August, 1985, but the disorder probably occurred earlier. During 1986, discussions with growers in southwestern Indiana revealed that symptoms identical to those associated with SDS were observed in 1981, and that, in 1979, several fields were affected with an unknown premature death of plants in patches.

Yield losses due to SDS range from 20 to 80% (240,241) or more, depending on variety and when the symptoms first appear. Reports of 40 to 60% yield losses are common. Appearance of the disorder at early pod fill is reportedly more damaging than its appearance at a later time (240). Yield reduction is reported to be due to pod abortion, lack of pod fill and low test weight (240,241).

SYMPTOMS OF SDS

SDS is most commonly found in fields with a high yield potential (Hirrel, M. C., and Rupe, J., Personal Comm., 240,241,453). SDS may be found in upland as well as river bottom fields, and apparently anywhere within a given field. In Indiana, symptoms of SDS frequently are more prevalent in those areas of a field with a slightly higher water holding capacity. SDS has been reported to occur more frequently in irrigated than in non-irrigated fields in Arkansas and Mississippi (Hirrel, M. C., and Rupe, J., Personal Comm.).

The symptoms of SDS may occur any time from late bloom through pod fill (Hirrel, M. C., and Rupe, J., Personal Comm., 60,238,240,453,564). The time of symptom expression appears to be related to (a) weather conditions (Hirrel, M. C., and Rupe, J., Personal Comm., 60,240,241,453,564), (b) variety (Hirrel, M. C., and Rupe, J., Personal Comm., 240,241,453), (c) maturity group of the variety (Hirrel, M. C., and Rupe, J., Personal Comm.), and (d) general growth conditions (Hirrel, M. C., and Rupe, J., Personal Comm., 241).

Symptoms begin to appear on apparently healthy plants. There are no obvious differences in plant height or other growth parameters between affected and healthy plants when the first symptom of SDS appears as an interveinal chlorosis (yellow blotches between the veins). The veins remain green. Middle to upper leaves frequently are affected first, although some varieties will first exhibit symptoms at the base of the plant while others may exhibit symptoms generally over the entire plant. The interveinal chlorosis increases from blotches to large irregular areas of interveinal tissue which quickly becomes necrotic. It is at this point that growers often first recognize the problem. Patches of plants with dying leaves become visible from a distance. As the symptoms of the disorder progress, the leaves of entire plants become affected. Defoliation may occur when SDS develops early enough in the growing season.

Throughout the development of the foliar symptoms, there is no downward bending of the petioles or other signs of wilting; the petioles remain erect. Defoliation occurs with leaf blades separating from the petioles, leaving the petioles attached to the stems. Stem and petiole epidermal tissues remain green for a period following defoliation. This pattern of defoliation was associated with SDS in Indiana during 1985 and 1986. In 1985, however, this same pattern was also observed in some varieties, primarily Century and a few Group III maturity varieties, in the central part of the state, but without other symptoms of SDS. Therefore, at this time, this defoliation pattern, by itself, is not considered diagnostic for SDS.

Pod abortion has been noted as one of the major yield loss components. The degree of abortion appears to be related to variety and time of symptom onset (Hirrel, M. C., and Rupe, J., Personal Comm., 240,241).

Close examination of the roots of plants just beginning to show interveinal chlorosis reveals near normal conditions. The lower portion of the tap root is the first to become discolored. There is no discoloration of the vascular tissue of the stem at this point. With time, the tap root becomes progressively discolored, and the lateral roots and nitrogen fixing nodules begin to deteriorate. A light milky-gray to milky-brown discoloration of the vascular tissues in the stem appears as the upper tap root becomes discolored. Usually, significant leaf necrosis and some defoliation are evident before stem discoloration occurs. Pith tissues of SDS-affected plants remain normal and white. Thus, the diagnostic symptom for SDS is the milky-tan to milky-gray discoloration of the cortical and vascular tissues of the stem with normal stem pith tissues. Discoloration of the stem may involve from half to three-fourths, or more, of the stem of severely affected plants. In addition to normal pith tissues, the epidermal tissues of SDS affected plants also appear normal.

Symptoms appear in somewhat circular to elongated patches in the field. The patches may vary in size from a single plant to large areas of an acre or more. Patches may coalesce resulting in large irregular areas of affected plants. The number of patches per field is highly variable, but there are generally multiple scattered patches. There have been only a couple of reports where SDS occurred generally over most of a field (Hirrel, M. C., and Rupe, J., Personal Comm.).

Plants growing under high temperature, drought, or other forms of stress do not develop symptoms of SDS. The length of time from symptom onset to plant death appears to be from 10 days to 3 weeks, although periods of up to 6 wk have been reported (Hirrel, M. C., and Rupe, J., Personal Comm., 240).

Discoloration of the cortical and vascular tissues may easily be confused with discoloration of the same tissues from Phytophthora root rot or with the symptoms of several other diseases. There are no signs of wilting in SDS affected plants while _Phytophthora_ infected plants wilt before yellowing and dying. SDS symptoms can easily be confused with the foliar symptoms of brown stem rot; however, with brown stem rot, the pith tissues are discolored, not the cortical tissues. In addition, Rhizoctonia root rot, charcoal root rot, stem canker, and other diseases may be confused with some of the symptoms of SDS. With each of these other diseases, the pathogen is readily identifiable or there are diagnostic symptoms that distinguish from SDS.

ROLE OF SCN IN SDS

The role of the soybean cyst nematode (SCN) in the development of SDS is not clear. Most states with SDS have reported that SCN was found in association with SDS when soil samples from affected fields were analyzed in the laboratory. One state, however, has reported that severe SDS was found in a field where SCN was not detected in a laboratory assay of the field soil (Hirrel, M. C., and Rupe, J., Personal Comm.). Preliminary variety testing data from Illinois and Kentucky indicated that 'Fayette' (resistant to SCN races 3 and 4) was moderately susceptible to SDS and more susceptible than some varieties without SCN resistance (Hirrel, M. C., and Rupe, J., Personal Comm.). Arkansas research suggests that, within maturity groups V and VI, there appears to be a trend toward SDS tolerance with SCN resistance, although the trend does not hold true with all varieties (241). In our August, 1985, investigations of SDS in southwestern Indiana, cysts of the nematode were not observed on roots of affected plants. However, laboratory analysis of the initial soil samples taken from affected areas were found to contain juveniles of the soybean cyst nematode. After the initial soil samples were analyzed, additional soil samples were then obtained from affected and non-affected areas of the same fields. In all cases, the nematode counts from affected areas were higher than from the non-affected areas (241). In our 1986 research, we found that cysts were not as prevalent on roots as would be expected with the high nematode populations that were in the research plots, and there was no correlation between SCN populations and the severity of SDS symptoms. However, SCN have been found in soil samples from all SDS affected areas investigated in Indiana to date.

WEATHER FACTORS RELATED TO SDS

Earlier reports indicated that SDS symptom appearance occurred after the movement of a major weather front brought cooler temperatures to an area, especially when the front was accompanied by rain and occurred at or near the flowering stage (Hirrel, M. C., and Rupe, J., Personal Comm., 240). In 1985, symptoms were first noted by farmers in southwestern Indiana during the first week of August. Temperature and rainfall data from Evansville, IN (National Weather Service) indicate that temperatures for the last two weeks of July and the first two weeks of August were normal; rainfall was just slightly below normal. The third week of August was significantly above normal for rainfall. During 1986, the average daily temperatures in our research plots were from 1 to 8 degrees below the 40 year average of 77.5 degrees F for 12 of the 14 days prior to the first SDS symptom development. Also in 1986, the research plot received 1.35 inches of rainfall five days before the first symptom of SDS developed. The 1986 weather patterns were close to those described above for SDS development, yet SDS was much more severe in the area in 1985 than it was in 1986.

CAUSE(S) OF SDS

The causal agent(s) of SDS has not been determined. Thousands of isolations have been made, but nothing consistent has been found. Electron microscopy studies showed bacteria plugging the vascular tissues (60), and a bacterium, tentatively identified as *Pseudomonas* sp., was isolated and produced symptoms when inoculated into healthy plants (241,564); however, symptom development was inconsistent. A fungus, identified as *Fusarium solani*, has been isolated, and research is continuing (431). In addition, other fungal pathogens such as *Diaporthe*, *Phomopsis*, *Macrophomina*, *Fusarium*, etc. have been isolated from SDS affected plants, but none have been demonstrated conclusively to be the causal agent of SDS.

Diagnostic techniques that are reliable for the detection of viral and mycoplasma pathogens have been negative to date. Results from tests for chloride toxicity (241) and other nutritional factors (Hirrel, M. C., and Rupe, J., Personal Comm.) suggest that these are not the causal agent of SDS.

Complexes of two or more interacting organisms, the production of toxins, or other factors can not be overlooked as possible causes of SDS. Research into these areas will continue.

The symptomatology, occurrence and pattern within a field, and the fact that methyl bromide-chloropicrin soil fumigation eliminates the occurrence of SDS symptoms (Scott, unpublished data) strongly suggest that SDS is of soil-borne, biotic origin.

CONTROL

Control measures for SDS are not defined at this time, yet, three practices are thought to be beneficial.

1. Crop rotation. While the total effect of crop rotation on SDS is unknown, this practice is valuable in the control of other soil-borne diseases. Also rotation will aid in reducing SCN populations. Preliminary research in Indiana suggests that SDS is more extensive in continuous soybeans than in a corn-soybean or corn-soybean-wheat rotation (Abney, T. S., Personal Comm.).

2. Tillage practices. Preliminary research data suggest that SDS is more prevalent in no-till than in chisel or conventional tillage (Abney, T. S., Personal Comm.).

3. Variety selection. In those areas where SDS occurs or is suspected, select varieties that performed well for the local area in those years when SDS was severe. The use of a SCN resistant variety should be strongly considered, although certain SCN resistant varieties appear to be susceptible to SDS.

4. Nematicides do not control SDS (Hirrel, M. C., and Rupe, J., Personal Comm.).

5. SDS does not appear to be seed-borne (Hirrel, M. C., and Rupe, J., Personal Comm., 240,241).

6. There are no economic chemical controls known for SDS. Methyl bromide-chloropicrin will control the appearance of symptoms, but the cost per acre is prohibitive.

PHYTOPHTHORA ROT OF SOYBEAN

A. F. Schmitthenner

The Ohio State University
Wooster, OH

Phytophthora rot of soybean (PRR), caused by *Phytophthora megasperma* f. sp. *glycinea* Kuan and Erwin (Pmg), is one of the most severe diseases of soybean in the northern Midwest (21,447) and is found throughout the soybean growing region. An estimated 10 million acres could be significantly damaged yearly by this disease. But, depending on the environmental conditions, only about one-third of this acreage is noticeably damaged in any one year. The greatest yearly damage measured in Ohio from *Phytophthora* has been about $50,000,000. The disease was first severe in the 1950's in the Maumee River basin of Northwest Ohio and Northeast Indiana. By the 1960's it was causing concern in Illinois, by the 1970's it was prevalent in Kansas, Iowa, and Wisconsin, and by the 1980's it was causing damage in Minnesota (276) and the Dakotas (153). Origin of the disease is not known, but the pathogen probably arose as a variant of native *P. megasperma* pre-existing throughout its range. Literature on *Pmg* is too voluminous to cite completely. Only recent reviews, the most recent papers on each subject, and little known papers are cited here; older literature can be accessed from these publications.

SYMPTOMS

Pmg can rot seed and seedlings at any stage of development (447). Rotting of the seedling at the crook stage is the most common damping-off symptom. Seedlings emerge, but fail to develop beyond the crook stage because the root system has been destroyed. Older seedlings wilt and die as their root systems are destroyed. The characteristic of Phytophthora damping-off, as opposed to other types of seedling diseases or herbicide damage, is that the root system is destroyed first, then the seedling wilts and dies. *Pmg* can cause lesions on, and rot, immature leaves (49).

Symptoms vary in older plants depending on the cultivar tolerance and plant age. In old seedlings (stages V2-V6) of low-tolerant (highly susceptible) plants, roots generally are completely rotted. A girdling, dark brown rot advances up the stem from the soil line. In older, low-tolerant plants (stages V7-R3) this advancing rot may extend as high as 10 nodes before the plants wilt. Prior to wilting, diseased plants with advancing lesions can be identified by a distinctive drooping or flagging of leaves. At this stage the tap rot is rotted throughout. Wilted plants die and dry up without loss of leaves.

In moderately-tolerant plants the characteristic symptom is root rot. Stem lesions are generally restricted to one side and are linear, chocolate brown, and somewhat sunken. Plants rarely wilt and die but are significantly stunted and may prematurely ripen. In highly-tolerant (slightly susceptible) plants, post-seedling damage is confined to the secondary root system and the tap root is not rotted. The plants appear healthy, but are significantly stunted when compared to disease-free plants (multirace resistant or Ridomil-treated plants).

This latter phase of Phytophthora root rot can be referred to as hidden damage. It also is characteristic of low- or moderate-tolerant plants that have escaped early-season disease but become infected later in the year.

DISEASE CYCLE

Pmg is a major cause of poor soybean stands in Ohio. It overwinters in soil as oospores (the sexual stage). Pmg cannot be detected in soil by a leaf baiting technique, after it has been frozen, dried, or stored at 3 C for an extended period of time (73,75). However, if such soil is incubated at a matric potential maintained between -100 and -300 mb for several weeks at 15-30 C before flooding, Pmg can be isolated. The number of leaf discs colonized depends on the time of incubation above 15 C and the actual incubation temperature. From this type of data it is assumed that Pmg is typical of other Phytophthora spp. in its requirements for sporangia formation and germination (125). Oospores germinate whenever moisture and temperature are suitable, forming sporangia that accumulate until the soil is flooded. Then, sporangia germinate indirectly forming up to 50 zoospores each. Zoospores swimming in the flood water are attracted to sites on roots and seed that are leaking nutrients. There, they encyst, germinate, and infect the host tissues. Leaf discs floating over soil containing sporangia can be infected within 90 minutes, and it is assumed that seed and root tissues can also be infected as rapidly. In soil that is not saturated, zoospores are not produced but infection may occur when a young root grows adjacent to a sporangium (21,280).

Stand loss from damping-off generally occurs in mid- to late-May plantings. Severe PRR, generally, results from infection of plants in the seedling stage. PRR is considered a simple interest disease. Even though secondary inoculum may cause some disease, the majority of infections probably result from primary inoculum (518). Late-season infections, generally, are restricted to the secondary roots but under certain conditions may cause severe damage. The best predictor for yield loss is disease incidence (number of yield-producing plants) at stage R7 (519). PRR also reduces plant height and affects other yield components. Severity of root rot is dependent on cultivar tolerance, Pmg races present, amount of inoculum in the soil, and the various environmental parameters that can be altered by cultural practices.

RESISTANCE

The first resistance to Pmg was found in the 1950s, at about the time the disease was identified. This resistance, labelled Rps, was incorporated into most of the northern cultivars and was satisfactory until 1971, when many fields of Harosoy 63 and Amsoy 71 were severely damaged in Ohio (447). A new race, called Race 3, was identified as a cause of this epiphytotic. Race 2 had previously been identified in the South. An intensive search was made for new sources of resistance and many were found. Now, there are seven different RPS genes known, with multiple alleles at both Rps_1 and Rps_3 loci (21,400). There also are a number of sources of resistance that have not been fully described. For example, Harosoy, originally thought to be a universal suscept, is now known to be resistant to several races described from the South. Relationship of the Harosoy gene (rps) to others is not yet known, so it is tentatively designated as Rpsh.

Resistance is evaluated by inoculating seedling hypocotyls or cotyledons with mycelium or zoospores of Pmg (365). The hypocotyl slit method is used most widely. If a plant is resistant, a slight lesion or hypersensitive reaction will occur (incompatible interaction). If a plant is susceptible, the hypocotyl or cotyledon will rot in 2 to 4 days (compatible interaction). Both temperature and tissue age are critical in evaluation of resistance. Resistance is not expressed at high temperature (99), and may be affected by temperatures as low as 28 C (271). Old tissues of susceptible plants express resistance similar to Rps resistance in young tissues (49,538). Rps resistance is race-specific. Each resistance gene will control some, but not all, races of Pmg Therefore when a "resistant" cultivar is attacked by a race to which it is not resistant, a susceptible reaction will occur (21,386,447).

Pmg races are morphologically similar, but have different cultivar host ranges. Discovery of new races has kept pace with incorporation of Rps resistance. In 1972 there were 3 races; by 1977, 9 races; by 1982, 20 races; and by 1986, 25 races (301). In addition, there are several reports of races not yet named that differ from the above (153,245,517,521). Since 1975, a uniform set of differentials has been used to identify races (21). Races have been numbered sequentially, depending on the date of discovery. There is concern that a better method should be adopted to name Pmg races; i.e., that each race be given a formula indicating its virulence or avirulence on specific Rps genes (521). A new set of differentials is being prepared at Harrow, Ontario, which will consist of each Rps gene or allele by itself in a Harosoy background. Once these differentials are available it should be possible to more accurately describe and separate old and new Pmg races.

Most resistant cultivars now have either Rpsl or Rpslc. This type of resistance is no longer acceptable in the northern Midwest because of the prevalence of Race 4. There are a limited number of cultivars with resistance to many races controlled by Rps_1^k (Williams 82 and Gnome 85), Rps_1^k + Rpsh (Century 84, Pella 86), Rps_1^b + Rps_3 (Keller) and Rps_1 + Rps_3 (Winchester). Unfortunately, race 25 is virulent to Rps_1^k, so this resistance may not be adequate for very long. Other two- and three-gene combinations are being incorporated for future release (21). These various multiple gene combinations should be usable for a longer period of time. Breeders probably can continue to pyramid genes for resistance in a limited number of cultivars as rapidly as new races are found. However, incorporating multiple genes for resistance into cultivars is very time-consuming. By the time resistance is incorporated, newer, higher-yielding cultivars will be available. The grower still will be faced with the decision of planting the highest yielding cultivar and risking disease, or planting a slightly lower yielding cultivar to stabilize yield at a sub-optimal level. A second disadvantage of resistance is that when root rot stress is completely removed, some cultivars grow so vigorously that lodging is abnormally severe and significantly reduces soybean yield.

MECHANISM OF RESISTANCE

Because of its unique race-resistance gene combination, the Pmg soybean system has been a popular model for studying the mechanism of resistance. The resistance reaction is typified by the rapid synthesis (54) and accumulation of post-infectional fungitoxic metabolites called phytoalexins (glyceollin isomers) (386,567). Glyceollins accumulate at the fungal-plant interface in a very short period of time (6-12 hr.) (201). Resistance is elicited by beta 1,3-glucan produced by Phytophthora and oligogalacturonides released by endopolygalacturonase from infected soybean cells acting synergistically (112). Elicitation of glyceollin has been demonstrated using both biotic and abiotic factors and the physiology and biochemistry of elicitation appears the same for both (567). It is probable that the action of elicitors is not variety specific, but, rather a part of a general defensive response of plants. Another type of response in soybean to Phytophthora can be elicited that involves an "ethylene" pathway and results in resistance without accumulation of glyceollin (Graham, unpublished). It is not known if this type of resistance is elicited by the pathogen under natural conditions.

The nature of susceptibility to Phytophthora in soybean is less well understood. One model is that virulent races are not recognized and, therefore, do not elicit the cascade of reactions resulting in resistance (31,54). This model is based on the gene-for-gene theory (31,272), and implies that: (1) resistance genes in plants and avirulence genes in pathogens are inherited as dominant characters; (2) avirulent races are specifically recognized by resistant cultivars; (3) specific recognition triggers a hypersensitive reaction and the accumulation of glyceollin; (4) virulent races are not recognized by susceptible cultivars and therefore do not trigger resistance; (5) avirulence is active, but virulence is a passive host response; (6) avirulence is physiologically dominant to virulence; (7) mixed virulent and avirulent races always will trigger an incompatible host response. The Pseudomonas syringe var. glycinea-soybean interaction appears to be a good example of this model (312).

Several aspects of the Pmg—soybean interaction do not appear to fit this model. After 10 years of intensive research, no race-specific elicitors have been found. Mixtures of virulent and avirulent races result in a compatible response (245, Schmitthenner, unpublished). Thus, virulence seems to be dominant to avirulence in Pmg. There is evidence for production for at least one race-specific suppressor of glyceollin (568), an alpha-mannan that appears to be localized at the hyphal tips of Pmg (219). In recent experiments, Clark and Breaux have established a race-specific "peroxide burst" and activation of an NADPH-dependent oxidase, after treatment with incompatible zoospores (unpublished). These types of reactions, called "prompt responses," are involved in race-specific suppression of the hypersensitive reaction in the Phytophthora infestans - potato system (121), which appears to fit the elicitor/specific suppressor model (31). This model proposes that resistance is due to a non-specific elicitor effecting cell death and the accumulation of phytoalexins, whereas, susceptibility is achieved by specific suppressors of resistance that prevent cell death or interfere with phytoalexin accumulation. Others have suggested that susceptibility may not involve lack of recognition but might be a more active process (69,205). There are a number of host-pathogen interactions where compatibility results from suppression of resistance (378). Farmer (150) has reported that one product of the PAL pathway, lignin, is suppressed by an elicitor from Pmg It is possible that the Pmg-soybean interaction also fits some kind of suppressor model, but more evidence of specific suppressors is needed.

Our knowledge of the genetics of virulence of Pmg is very primitive and insufficient to support the gene-for-gene theory of disease or either of the above resistance models. Pmg is diploid in the vegetative stage, as are all other Phytophthora spp. Erwin (148) concluded that mycelium of Phytophthora probably is heterozygous. Single oospore progeny were probably more variable than single zoospore progeny because of genetic recombination following meiosis in oospores. Single zoospore progeny were found more variable than mass transfers by Rutherford et al. (435). They also found three distinct virulence patterns among second-generation single zoospore isolates of race 6. From an isolate of race 3 that was no longer virulent, a single zoospore with a race 1 reaction was obtained among the first generation of zoospores, and isolates with a race 3 reaction were among the second generation. They considered the variability to be too great to be accounted for by conventional mitotic mechanisms, such as mutation or crossing-over, and suggested that cytoplasmic genes might be involved. They suggested such random generation of races was consistent with evidence by Hobe (245) that a wide range of new races may be present in the soil.

Pmg genetics is difficult to study because the fungus is homothallic, the oospores do not germinate consistently (260), and oospore germination may be under genetic control (434). Some oospore germination does occur (195,260,280), but parasexual genetics has been most useful with Pmg. Long et al. obtained pure breeding Pmg after five generations of selecting single oospores (313). Using these pure lines they constructed heterocaryons from drug-resistant and auxitrophic strains. Segregation of the selfed heterocaryons was consistent with segregation expected in a homothallic, diploid fungus. Layton and Kuhn (unpublished) constructed metalaxyl and 6-fluorotryptophan resistant mutants of races 1 and 3 and heterocaryons were formed by protoplast fusion. Race 1 reaction (avirulence) was phenotypically dominant to the race 3 reaction (virulence), which they considered supported the gene-for-gene hypothesis. Evidence for somatic recombination was obtained in the progeny of the heterocaryon. More information of this nature on the inheritance of virulence is needed before the genetics of the interaction can be understood. It is concluded that in spite of the voluminous research on the Pmg-soybean interaction, the nature of resistance and susceptibility is still not clearly understood.

FIELD RESISTANCE, TOLERANCE, RATE-REDUCING RESISTANCE

Not all susceptible cultivars are equally damaged by virulent races of Pmg, Some appear resistant, yet they are susceptible when hypocotyl-inoculated with virulent isolates. These cultivars can be damaged in the seedling stage, but their tap roots are not rapidly colonized by virulent Pmg races. They are considered to have field resistance, tolerance, or rate-reducing resistance, which in this paper are considered synonymous (447). Tolerance appears to be race non-specific (effective against all races). Cultivars that are severely damaged are referred to as low-tolerant, moderately damaged lines as moderate-tolerant, and slightly damaged genotypes as high-tolerant. Tolerance can be evaluated by measuring plant loss and vigor in the field, by inoculating the roots of seedlings in an inoculum layer test, by a modified slant board test, or by a number of other laboratory and greenhouse procedures (372,529). Generally, there is good agreement between laboratory tests and field response. High tolerance (rate-reducing resistance) involves reduced colonization of the tap root (372,520). The mechanism of reduced tap root colonization has not been elucidated, but glyceollin accumulation is not involved (373). Apparently, reduced tap root colonization has a different biochemical basis than race-specific (Rps) resistance.

There are many cultivars with excellent tolerance to Pmg, but relatively few that combine high yield and tolerance (46). Heritability of tolerance is relative high (530), and genetic improvement can be achieved through recurrent selection (531). There is some evidence for intracultivar variability in tolerance, so improvement possibly could be achieved by selection within adapted, high-yielding cultivars (373). High-tolerant cultivars can be used to reduce Pmg damage, but will not completely control the disease under severe conditions. The weaknesses of tolerance are: (1) it generally is not effective in the seedling stage; (2) hidden damage may be present, with more yield loss occurring than is evident; (3) it appears to involve several quantitative genes and thus is difficult to combine with other quantitatively inherited characteristics such as yield; and (4) it may be day-length sensitive; i.e., Group 3 and 4 cultivars are more tolerant in Ohio than Group 2 that are considered high-tolerant north of Ohio, compared to lesser tolerant Group 1 cultivars. South of Ohio, Group 3 and 4 cultivars are not considered to have good tolerance. With all of these weaknesses, high tolerance still is a good background on which to develop both integrated control and chemical control of Phytophthora root rot.

EFFECT OF CULTURAL FACTORS ON PHYTOPHTHORA ROOT ROT

Compaction

Phytophthora root rot always is more severe in the headlands or other locations in the field that receive heavy equipment traffic. Gray and Pope (195) demonstrated that increased bulk density produced by driving a tractor over disced soil resulted in increased _Pmg_, with reduction in number of yielding plants per row and reduced yield compared to non-compacted controls in two years. No mechanism for the increased root rot was suggested. Bulk density and severity of Phytophthora root rot may be reduced by spring plowing. Effects of tillage systems need to be further explored.

Tillage Intensity

There is increasing evidence that Phytophthora root rot is more severe in soybeans produced in no-till (conservation tillage) than in complete tillage (plowed, field cultivated or disced before planting) cropping systems. In Ohio, yield of the high-tolerant cultivar Asgrow 3127 was 19% less in no-till than in completely tilled plots (fall plowed, field cultivated before planting) (450). Low-tolerant cultivars also respond to tillage in some years (Schmitthenner and Van Doren, unpublished). In field tests, yields of Asgrow 3127 can be increased 20% by application of a Ridomil soil treatment in no-till production (Schmitthenner, unpublished). In Indiana tests (452), 54% PRR was reported in no-till, 9.7% in chisel-plowed, and 1.0% in moldboard-plowed soil. Tillage intensity also has been related to severity of PRR in Iowa (499). The mechanism whereby Phytophthora root rot is enhanced by no-till has not been determined. Possible mechanisms are effects on drainage, bulk density, and inoculum distribution (450). Control of _Pmg_ needs serious consideration in no-till production systems.

Drainage

Soil moisture is a one of the most critical factors affecting severity of Phytophthora root rot (120,447). Saturated soil is essential for indirect germination of sporangia (zoospore production) and dispersal of zoospores to infection sites on roots (125). As little as 1.5 hr of soil flooding may result in significant trapping of zoospores by leaf discs (73,75). Observations in Ohio (Schmitthenner, unpublished) have indicated that soybeans that are not flooded soon after germination may escape severe _Pmg_ damage. Drainage has a greater impact on low- than on high-tolerant cultivars (450). Generally, _Pmg_ is less severe in well-drained soils, but the relationship of soil type to time of soil saturation and Phytophthora root rot severity has not been quantitated. Drainage can be improved by tiling. However, even tiling will not prevent damage from excessive rainfall (450). The interaction of length of saturation period, time of saturation with respect to soybean development, cultivar tolerance, and severity of Phytophthora root rot needs to be established. Also, the role of soil saturation in enhanced severity of _Pmg_ in no-till cropping systems has not been explored.

Crop Rotation

Continuous cropping (monoculture) may increase severity of Phytophthora root rot (18). In a 5-year study, plant loss in Harosoy (low-tolerant) increased more than in Coles (high-tolerant) which was more than in Corsoy 79 (resistant). However, increases in yield losses were not clear cut because of high year-to-year variability. It was suggested that inoculum may have increased over the 5-year period. In concurrent laboratory studies, more oospores were produced in roots in a high-tolerant than in low-tolerant or resistant cultivars. The relationship of oospore inoculum density to disease severity has not been determined because of difficulty in isolating and germinating oospores from soil (18,260). Active Pmg inoculum can be assessed by a leaf disc bioassay (73,75). Staining and counting viable spores in soil (102) or antibody techniques to quantitate Pmg biomass may prove useful for relating Pmg population density to disease severity.

There is less evidence that crop rotation will reduce Pmg severity. In Ohio, rotation with corn had a minor effect on Phytophthora root rot and, then, only in combination with seed treatment and ridge tillage (450). In further work (Schmitthenner and Van Doren, unpublished), a 6% increase in yield could be assigned to rotation in a multifactor experiment in Phytophthora-infested soil.

There is some interest in Phytophthora decline. Kilen (unpublished) has reported that symptoms of Phytophthora root rot have become less severe in susceptible cultivars in experimental monocropped (30 yr) fields planted in alternate years to resistant cultivars. Phytophthora decline also has been suggested in Iowa (498). In Ohio, Phytophthora tolerance has been evaluated in the same field since 1972. Disease severity has fluctuated with weather conditions, but has not declined over the 14-year period. The impact of moisture, tillage, and cultivar tolerance need to be evaluated before Phytophthora decline can be verified.

Chemical and Organic Fertilization

High fertility has been shown to increase severity of Phytophthora root rot (120). In growth chamber tests (73), chloride was found to be the critical factor involved. Root rot was increased by potassium chloride but not potassium sulfate. Other chloride salts increased the percentage of infected plants. In preliminary tests in Ohio (Schmitthenner, unpublished), 40 metric tons/hectare of composted municipal sludge (CMS) significantly decreased yield in Phytophthora-infested soil when added immediately before planting. When added 3 mo. prior to planting, yields were significantly increased by the same CMS application. The yield-reducing effect of CMS could be mimicked by application of its salt equivalent in the form of sodium chloride and calcium phosphate. Also in Ohio, it has been observed that parts of fields that consistently receive the most manure have the most severe root rot. Chloride also may be involved in this effect. The significance of chloride in severity of Phytophthora root rot needs further study. To avoid possible interaction, potassium chloride fertilizer, CMS, and manures should be applied to soil far enough ahead of planting to allow for chloride to leach out of the root zone.

Herbicide Interactions

Trifluralin, a dinitroanaline herbicide, has been shown to increase severity of PRR (124). The effects of other dinitroanaline herbicides are not known. A second herbicide that increases PRR is 2,4-DB (537), which is commonly used as an emergency treatment for various broadleaved weeds. Plants with Pmg-infected roots will develop stem lesions after treatment with 2,4-DB, but its effects on enhanced yield loss have not been thoroughly evaluated. Glyphosate, which may be applied with a wick to weeds in soybean fields, reduces resistance and glyceollin accumulation (273). No other herbicides have been reported to increase PRR damage. However, in Ohio, metribuzin damage is frequently confused with Pmg damage because they both occur most frequently in the headlands and in wet spots. Pmg kills the roots first, then the tops wilt. Metribuzin kills the tops first, then the roots rot. Metribuzin never causes damage prior to the first trifoliolate stage.

Biological Control

There is continued interest in biological control of Phytophthora root rot. Two approaches have been made. At Michigan State, emphasis has been on destruction of oospore inoculum by hyperparasitism (162,486). Many hyperparasites have been found effective in greenhouse tests, but they need to be tested extensively in the field to determine if any can offer a practical control of PRR.

The second approach has been to protect seed and seedlings with seed dressings of antagonistic bacteria. One potential bacterial biocontrol agent has a patent pending (Paxton, J., Illinois, personal communication). In extensive tests with native rhizobacteria, none consistently controlled Phytophthora root rot under high inoculum conditions (306). It is concluded that no viable biocontrol for Pmg is yet available.

FUNGICIDES FOR PHYTOPHTHORA ROOT ROT

The first *Phytophthora*—specific fungicides were found in the late 1970's. Pyroxyfur (Grandstand) was effective as a seed treatment, particularly on high-tolerant plants (302). Pyroxyfur has never been marketed because soybean seed treatment as a sole market was considered too small for its development.

Metalaxyl, developed by Ciba-Geigy for downy mildew diseases and black shank of tobacco, is even more effective against Pmg than pyroxyfur. It is used both as a seed (Apron) and a soil (Ridomil) treatment. As a seed treatment, metalaxyl provides adequate control on high-tolerant but not on low-tolerant cultivars (450). Seed treatment is not effective for control of root rot. Apparently, most of the metalaxyl applied to seed is translocated to the foliage (200), and too little is translocated to the roots to provide effective control. Metalaxyl seed treatment also may be phytotoxic (reduce yield) if applied at too-high rates, or if treated seed is germinated under certain adverse conditions (Schmitthenner, unpublished). Advantages of Apron are that it provides protection against damping-off at a modest cost (about $2.50/25 kg seed). Disadvantages are that (1) it does not prevent root colonization, (2) it adds to seed costs and treated seed is not competitive with untreated seed on the market and (3) seed companies are reluctant to risk an unsold inventory of treated seed that must be destroyed.

Soil treatment with metalaxyl (Ridomil) is even more effective than Apron seed treatment for control of Phytophthora root rot (450). Control is more effective on high- than on low-tolerant cultivars (450,518). Rates commonly used in experimental work, 1.12 kg a.i./ha (450,518), are too expensive ($148/ha) for soybean growers. Research in Ohio (Schmitthenner, unpublished) has demonstrated that lower rates (0.28 kg a.i./ha) are almost as effective if applied in the root zone of high-tolerant cultivars, either as a granule in the seed furrow or as a narrow spray band (5 cm) over the seed furrow ahead of the press wheel. These rates correspond to the minimal rates on the Ridomil label. Higher rates are necessary for good control if Ridomil is broadcast (0.56 kg a.i./ha, cost $74.00/ha) for Pmg control in drilled soybeans. This minimal broadcast rate will not control Pmg on low-tolerant cultivars. More research is needed to establish the minimum effective rate of Ridomil needed for good control when applied in the root zone in rows spacings of 35 and 17.5 cm. Cheap delivery systems for spraying in furrow could be adapted for these row spacings. Ridomil always will be too expensive for adequate control on low-tolerant cultivars under severe disease conditions.

Other acylalanine fungicides tested (oxydixyl and benalaxyl) have not been as effective as metalaxyl for PRR control (Schmitthenner, unpublished). Metalaxyl is a very active compound. LD90 rates in agar for inhibition of mycelium growth of *Pmg* are 150 ppb (101). Broadcast rates in soil (based on 15 cm depth) of 250 ppb (0.56 kg a.i./ha) are effective only with high-tolerant cultivars (Schmitthenner, unpublished). If metalaxyl is taken up equally well from soil by low- and high-tolerant cultivars, control failure with low-tolerant cultivars is puzzling. It is assumed that metalaxyl inhibits colonization of root tissue as the fungus moves through the roots to the top of the plant. It has been suggested that metalaxyl may elicit resistance response in susceptible plants. A more simple explanation for this differential control is that less metalaxyl is required in high-tolerant cultivars where colonization is already reduced by quantitative genes - an additive effect. Data on actual amounts of metalaxyl in root tissue are needed before we can determine if it acts synergistically or additively in high-tolerant cultivars.

Concern has been expressed that continued use of Ridomil in the soil will result in selection of Ridomil-resistant isolates. Although resistance to Ridomil can be induced in *Pmg* using chemical mutagens (111, Layton, A., and Kuhn, D, unpublished), there are no reports yet of naturally-occurring Ridomil-resistant isolates. In Ohio, Ridomil has been used in two fields for seven consecutive years and is as effective now as it was the first year. Soil application of Ridomil should be effective for many years.

CONTROL

Resistance

Cultivars need to be resistant to races 1, 3, 4, 5, and 7 to provide adequate control in the North Central Region. Cultivars with either the single gene Rps_1^k (Williams 82 and Gnome 85) or combinations Rps^k and $Rpsh^1$ (Century 84 and Pella 86) and of Rps_1^b or Rps_1^c with Rps_3 (Keller and Winchester) will provide control until additional virulent races develop.

Integrated Control

Combine high-tolerant cultivars, good drainage, complete tillage, rotation with corn, and Apron seed treatment for damping-off control. This option is not open for no-till systems because of the requirement of complete tillage to minimize root rot damage.

Metalaxyl Seed Treatment (Apron)

Metalaxyl seed treatment is effective on high-tolerant cultivars for control of damping-off, not root rot. Combination with Captan is better than with Vitavax 200 or PCNB. Dry hopper-box formulations are better than liquid formulations, probably because some of the metalaxyl is deposited into the soil at planting.

Metalaxyl Soil Treatment (Ridomil)

1. Minimum labelled rates (0.28 kg/ha) of metalaxyl as a soil treatment are adequate for control on high-tolerant cultivars in 38- or 76-cm spaced rows. Higher rates (1.12 kg/ha) may be necessary for good control on low-tolerant cultivars. Apron seed treatment will not improve performance of soybeans in Ridomil-treated soil.

2. Minimum labelled broadcast rates are suitable for control in drilled high-tolerant soybeans, but may be too expensive. Minimum labelled in-furrow rates will provide adequate control if applied in the root zone on cultivars with the highest level of *Pmg* tolerance. In-furrow liquid formulations should be applied as a narrow band (2-5 cm) sprayed over the furrow right behind the double disc openers.

THE SOYBEAN CYST NEMATODE

D. I. Edwards

Research Plant Pathologist, USDA, ARS,
and Associate Professor of Nematology
Department of Plant Pathology
University of Illinois
Urbana-Champaign, IL 61801

The soybean cyst nematode, *Heterodera glycines* Ichinohe (SCN), is an extremely devastating pest of soybeans. Since it was first discovered in the United States in 1954, infestations have been detected in 25 states (Table 1) involving most of our major soybean-producing areas. In 1979 loss due to SCN was estimated at 1.5 million metric tons (approximately 56 million bushels) (61). More recently, the Disease Loss Estimate Committee of the Southern Soybean Disease Workers estimated loss of soybean yields in 1984 due to SCN to average 2.48% for 16 southern states (361). The estimated losses for individual southern states ranged from none (Texas) to 4.9% (North Carolina).

In the North Central Region of the United States, nine states have infestations of SCN, with the most recent reports coming from Kansas (467) and Nebraska (D. S. Wysong, University of Nebraska, personal communication). The most extensive infestations occur in Illinois and Missouri, where combined annual losses have exceeded 680 thousand metric tons (61,139). In Illinois, the leading soybean-producing state, new county infestations of SCN have been found almost on a yearly basis since its first detection in 1959. As of the 1986 growing season, 81 of the 102 Illinois counties have infestations, with the most severe and extensive disease being in the southern half of the State. Increases in infested acreage have also occurred in Indiana, Iowa, Minnesota, and Wisconsin over the past five years. The author estimated loss of soybean yields in 1986 due to SCN to average 2.0% for the eight northern states designated in Table 1. The estimated losses for individual states ranged from a trace (Nebraska) to 7.0% (Illinois).

Cysts of SCN occur throughout the root zone in infested soil and can easily be moved with soil by activities of man and natural agents. In Illinois, cysts have been found in soil on farm implements, land grading equipment, vehicles, tools, and shoes. Nursery stock, vegetable transplants, and root crops may carry cysts in adhering soil, even though the plants themselves are not hosts of SCN. Hay, straw, grain, or seed crops that carry soil particles or peds may also serve as carriers. Basically, anything that moves through an infested field in contact with the soil is capable of picking up and transporting cysts either locally or over long distances (342). Natural agents such as wind, surface water, and wildlife can spread cysts into non-infected areas. Three species of blackbirds have been found to harbor living cyst nematodes (145). These birds may spread SCN from field to field or over long distances by their migration. Because of these natural factors, state and federal quarantines have not been effective in preventing the spread of SCN.

Table 1. States with recognized infestations of the soybean cyst nematode.

State	Year Discovered	State	Year Discovered
North Carolina	1954	Alabama	1973
Missouri [1]	1956	Georgia	1976
Tennessee	1956	Texas	1976
Arkansas	1957	Delaware	1978
Kentucky	1957	Maryland	1978
Mississippi	1957	Iowa	1978
Oklahoma	1957	Minnesota	1978
Virginia	1958	Ohio	1980
Illinois	1959	Wisconsin	1980
Florida	1967	New Jersey	1984
Louisiana	1967	Kansas	1985
Indiana	1968	Nebraska	1986
South Carolina	1971		

[1]States underline are considered to be in the North Central Region of the U.S.

SYMPTOMS AND IDENTIFICATION

Symptoms of SCN damage are not specific enough to allow positive identification because they can be confused with other crop production problems such as nutrient deficiencies, injury from agricultural chemicals, and other soybean disorders. However, some symptoms are highly suggestive of SCN infection. Plants in heavily-infested areas may be chlorotic or stunted in appearance. These symptoms may be greatly accentuated by low soil fertility or droughty conditions. Infested portions of a field may be oval to elliptical in outline, with the most severe damage, such as stunting and yellowing, occurring in the center of these areas. Frequent cropping of susceptible soybeans will expand the affected area in the direction of the tillage.

Because other conditions may cause similar symptoms, positive identification cannot be made on this basis alone. To correctly diagnose SCN infestation, it is necessary to see and identify the nematode in association with injured plants. Consequently, nematodes must be recovered from soil or plant roots and, because of their small size, be identified under a microscope or, in the case of root examinations made in the field, with a hand lens.

Generally, by four to six weeks after planting, field diagnosis can be made by digging plants from the margins of suspected infested areas and gently washing or tapping the soil from the roots. With a hand lens, or perhaps with the unaided eye, white to yellow, lemon-shaped females can be seen on the soybean roots. The absence of visible white to yellow females does not rule out SCN in the problem area. The absence of the nematode is not uncommon where soils are extremely wet or dry, or when root deterioration occurs from fungal infections or normal plant senescence. In such cases, soil samples should be collected and submitted for analysis (342).

Diagnostic soil sampling, used to determine if SCN is present in a field, can be done any time of the year, providing that the sampling tool for collecting soil can be inserted into the soil to the proper depth. Diagnostic information is usually gathered during the growing season when no remedial management practices are available to the grower (343).

Predictive soil sampling and analysis for SCN population levels is intended to provide timely information for growers, especially those considering planting soybeans the next growing season. Soil samples can be collected in the fall after harvest or in the early spring. Analysis of a fall collection may be more beneficial because it allows the grower more time to make planting decisions such as selecting and purchasing of soybean cultivars (susceptible vs. resistant) or changing the crop rotation sequence.

Soil samples suitable for predictive assays can be adequately obtained by taking 10 to 20 subsamples in a zig-zag pattern through each 5-acre area to be sampled (Fig. 1). Using a narrow-bladed shovel, a garden trowel, or soil probe, soil samples should be collected as close to the old crop row as possible to a depth of 6 inches, discarding the top one inch of the soil. Subsamples should be mixed and placed in a plastic bag and sealed. Soil samples for nematode analysis must be handled differently than those taken for fertilizer recommendations. To be recovered and identified, nematodes must reach the laboratory alive. Therefore, they should not be allowed to dry out or be exposed to temperatures above 32 C. Samples should be delivered or mailed as soon as possible after collection.

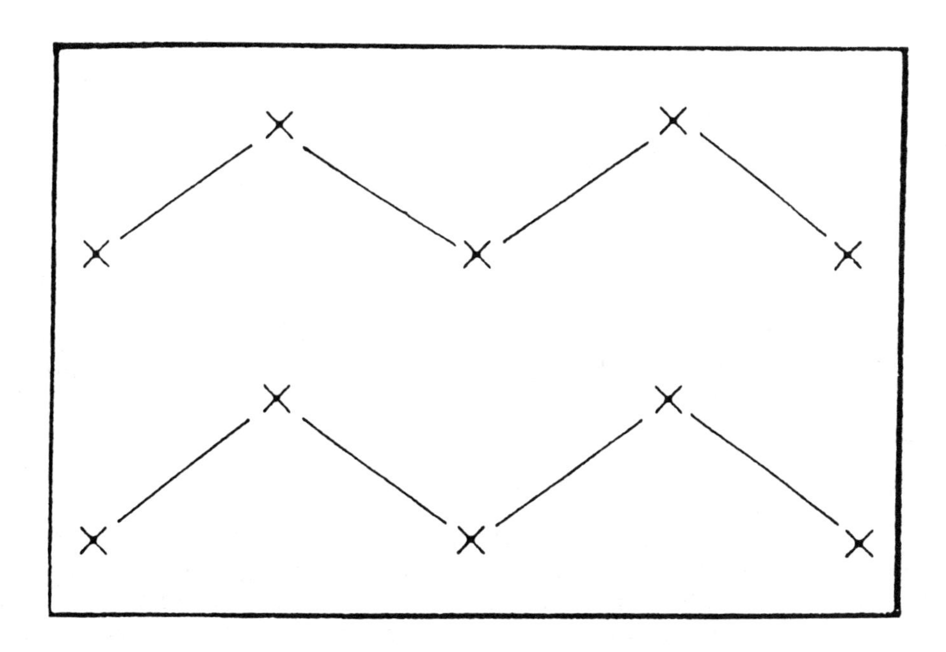

Fig. 1. Pattern for predictive SCN sampling in a 5-acre area. Each x=1 subsample, 10 subsamples minimum (343).

Damage threshold levels, which refer to the number of nematodes required in a given volume of soil (or weight of roots) to cause economic losses, have not been determined for most plant-parasitic nematode species, including SCN. In Illinois, a soil population below 20 viable eggs and larvae per 100 cc of soil is considered to be the population at which a susceptible variety can be grown without significant damage (342). This recommendation generally is applicable to low organic soils of southern Illinois. Studies are now underway to determine economic threshold levels for the higher organic soils found in central and northern Illinois.

SCN RACES

Five races of SCN (designated 1 through 5) have been identified and are characterized by their relative ability to reproduce on key soybean varieties (Table 2) (185,253). All five races of the nematode have been found in the United States, but races 3 and 4 appear to be most common. In the northern states, race 3 appears to be more prevalent, followed by race 4. Races 1 and 5 have been detected in Illinois but only in isolated cases (342). Race 5 also occurs in Minnesota (314,478).

Table 2. Characterization of the known races of the soybean cyst nematode in the United States (185,253).

	Reproduction on Key Varieties				
Race	Pickett[a]	Peking	PI88788	PI90763	Lee[b]
1	No[c]	No	Yes	No	Yes
2	Yes	Yes	Yes	No	Yes
3	No	No	No	No	Yes
4	Yes	Yes	Yes	Yes	Yes
5	Yes	No	Yes	No	Yes

[a] Varieties developed from crosses with Peking (as was Pickett) appear to have a similar response.

[b] Essex has also been used (after MacDonald et al. [314])

[c] No = number of white females < 10% of the number on Lee; Yes = number of white females ≥ 10% of the number on Lee (after Golden et al. [185]).

Resistant varieties are an excellent tool for use in SCN-control programs. However, the ineffectiveness of some resistant cultivars in reducing SCN losses has indicated much greater genetic variability in populations of this nematode and suggests the existence of more than five races. Workers in Arkansas (412) have separated 38 SCN populations into 25 distinct groups by determining their reaction on 13 soybean genotypes. This genetic variability of SCN populations is also indicated for the northern United States. For example, Illinois field collections of SCN do not always fit into the recommended race scheme shown in Table 2. This genetic variability is also exemplified by the failure of some resistant varieties shortly after their release for use by growers. This extremely wide range of genetic variation strongly points out the need to utilize control practices which minimize selection pressure on SCN populations (51,342).

CONTROL APPROACHES FOR SCN

Resistant Cultivars

Progress has been made in developing cultivars resistant to SCN. Resistance to races 1 and 3 has been primarily derived from the black-seeded cultivar Peking and incorporated into public cultivars with acceptable agronomic characters. Among these, four are adaptable to areas of the northern United States. These include CN210 (maturity group II), CN290 (II), Custer (IV), and Franklin (IV). Custer and Franklin are presently not grown on large acreage because they have been replaced by other genotypes with superior agronomic characters and the added resistance to race 4 of SCN.

In 1977, Bedford, the first cultivar resistant to race 4, was released in the southern United States. The resistance in Bedford was obtained from the black-seeded cultivar PI88788, which was crossed with Forrest. Like Forrest, Bedford is in maturity group V and is not adaptable to most soybean producing areas in the northern United States. However, two public cultivars, Fayette (III) and Cartter (III) with resistance derived from PI88788, have been released in 1981 and 1986, respectively. Fayette and Cartter are similar in disease reaction with moderately good resistance to races 3 and 4 of SCN, bacterial pustule, and several races of downy mildew. Cartter is approximately eight days earlier than Fayette which will extend its usefulness to more northern areas of the Midwest.

In recent years, some private breeders have made strides in developing soybean cultivars resistant to races 1, 3, and 4 of SCN. However, as with the public releases, the maturity groups of the private lines generally range from III through VI. There is now a need for SCN-resistant cultivars in maturity groups 0 through 2 for growers in northern Illinois as well as for most soybean producing areas of Iowa, Minnesota, and Wisconsin.

Crop Rotation

Crop rotation has proven to be an effective control measure for SCN. However, the time interval between soybean crops depends on infestation levels. Growing soybeans every other year in rotation with corn is a cropping sequence desired by many growers in the Midwest. In Illinois, this rotation has allowed SCN population levels to increase, although not as rapidly as with continuous soybeans.

To reduce SCN populations, a nonhost crop should be planted for two, or preferably three, years between soybean crops. Popular nonhost crops for the Midwest include corn, small grains, and alfalfa. If a grower prefers to plant soybean frequently in a rotation, then resistant cultivars can be included providing the race of the nematode present in the field is known. Such a rotation scheme is recommended in Illinois (342) and the procedures are as follows: Year 1, the first year after the identification of the nematode problem, plant a nonhost crop; Year 2, plant an adapted SCN-resistant cultivar; Year 3, return to a nonhost crop; Year 4, plant a high-yielding susceptible cultivar providing a soil analysis shows SCN populations are below the economic threshold level; and Year 5, repeat the rotation. The use of a sound rotation should not be underestimated. Higher yields will be achieved on all crops grown in the rotation than if any one crop is planted continuously. In the case of SCN, the use of a rotation involving nonhost crops and soybeans, both resistant and susceptible, will minimize the selection pressure on nematode populations and may prolong the useful life of resistant cultivars.

Chemical Control

In areas where resistant cultivars are unavailable, growers may want to consider using nematicides for the control of SCN. At present, compounds registered for control of SCN are aldicarb 15G, carbofuran 15G, ethoprop 15G, and phenamiphos 15G and 3E. Fensulfothion 15G is labeled for control of soybean nematodes but is not recommended for SCN. These chemicals are generally applied in 6- to 15-inch bands over the row at planting and incorporated to a depth of 2 to 6 inches. Carbofuran 15G and aldicarb 15G are also labeled for in-furrow application. Although all these nematicides offer some degree of control, aldicarb 15G has provided the most consistent yield response in tests conducted in Illinois (139,342). However, nematicides have not been used extensively because of their cost and the ability of resistant varieties to produce yields as large as those from soybeans treated with the most effective nematicide.

Biological Control

Parasites and predators have been observed for many years in association with naturally occurring nematode populations. Included among these potential biological control agents (BCA's) are protozoans, tardigrades, collembola, predaceous nematodes, a rickettsia-like organism, bacteria, viruses, and nematode-destroying fungi. Most studies to date have involved isolation, identification, and culturing of some of these organisms with limited evaluations of candidate BCA's in laboratory, greenhouse, and microplot experiments. Beyond this, there is no information on the effectiveness of these organisms to control nematodes in agricultural systems used by growers.

Since the 1970's, the study of nematode-destroying fungi as BCA's for cyst nematodes (_Heterodera_ spp.) has greatly increased. An extensive literature review covering this progress is well documented in a doctoral thesis recently completed at the University of Illinois (76). In the Illinois study, 69 species of fungi were identified from cysts collected from two SCN-infested fields during 1983-1985. One field used in this study had an apparently suppressed SCN population which did not increase during the three year period. Eighty percent of the cysts in the suppressed population were colonized by fungi and had a more relatively diverse fungal composition than cysts in a field where the SCN population continued to increase. Results strongly indicate that one or more of these fungi are involved in biologically suppressing SCN (76,77). Eight species of fungi were found to be the predominant colonizers of cyst and these will be evaluated individually as potential BCA's.

SOYBEAN NEMATODES IN THE NORTH CENTRAL UNITED STATES

T. L. Niblack

Department of Plant Pathology
Iowa State University
Ames, IA

The objective of this paper is to discuss nematode pathogens of soybean, other than the soybean cyst nematode (Heterodera glycines Ichinohe), that occur in the north central United States. This presents something of a problem, because those species, excluding H glycines, that occur in the north central states and for which we have convincing evidence of pathogenicity, are few. This does not mean that other nematode pathogens of soybean do not occur. One root-knot nematode (Meloidogyne sp.) is found in northern soybean-growing regions, as are some pathogenic lesion nematodes (Pratylenchus spp.). However, of the over 100 species of plant parasitic nematodes associated with soybean in the field (446), few have been studied in the North Central Region.

There is difficulty involved in identifying a problem with most soybean-parasitic nematodes. The usual above-ground symptoms produced by nematode parasitism are non-specific and often subtle, and there are generally no gross root symptoms such as the galls induced by root-knot nematodes, or recognizable signs such as the white females of H. glycines. Evaluation of the effect of nematode parasitism in greenhouse studies may be of little value, as soybean tends to be more tolerant of nematodes in this environment. Furthermore, the usual plant response data collected in greenhouse studies, such as root and shoot weights, may have little relationship to seed yields.

Most of the known important nematode pathogens of soybean are restricted in distribution to the southern and eastern states (369,446). Schmitt and Noel (446) produced an excellent review of those pathogens. The information available on nematode-soybean relationships in the north central states is reviewed herein, with more detailed information given for the better documented pathogens.

ROOT-KNOT NEMATODE

Meloidogyne hapla Chitwood, the northern root-knot nematode, is the common root-knot nematode species known to parasitize soybean in the north central states (369). Little is known of the specific M. hapla-soybean relationship. Most investigations of root-knot nematode-soybean interactions have been with the southern root-knot nematode, M. incognita (Kofoid and White) Chitwood which is a common and severe soybean pathogen in the southern and eastern states. M. hapla can be distinguished from other Meloidogyne spp. on the bases of host range, female perineal patterns, or male morphology (140,500). Furthermore, parasitism by M. hapla often results in a characteristic gall with one or more lateral roots proliferating from the gall.

Gall formation by soybean roots begins when the second-stage juvenile (J2) penetrates the root and initiates giant cell formation in the protostele. Giant cells are multinucleate transfer cells that form in a cluster about the nematode's head and on which the nematode feeds. Adjacent tissues undergo hypertrophy and hyperplasia.

M. hapla can also damage soybean indirectly through interaction with other organisms. Taylor and Wyllie (502) found an increase in pre-emergence damping-off caused by *Rhizoctonia solani* in the presence of *M. hapla*, and an additive interaction between *M. hapla* and *Phytophthora megasperma* var. *glycinea* (561). Soybean nodule formation, however, was stimulated in the presence of *M. hapla* in a greenhouse study (252).

Once a feeding site has been established, the small (0.4-0.5 mm), slender J2 become sedentary and begin to swell. Swelling increases as juveniles molt three times to become adults. Adult females are saccate and immobile within the root; males leave the root, do not feed, and are not involved in reproduction. Females produce a gelatinous material into which eggs are deposited at the root surface. Eggs may overwinter, or J2 will hatch in about 40 days or more depending on conditions.

There is no information about the specific relationship of *M. hapla* and soybean under field conditions, but in general soybean yield losses due to root knot are inversely related to J2 densities at or before planting (446). Development of predictive soil assays necessary for control recommendations in the north central states will require more information. Breeding soybeans for root-knot resistance has been directed primarily at *M. incognita*, but resistance to *M. hapla* is available (10). Reducing the *M. hapla* densities in a field to below damaging levels can probably be accomplished by rotation with corn, as corn is not a host for this species.

LESION NEMATODES

Twenty years ago, Ferris and Bernard (157) observed that only limited information was available on the effects of common *Pratylenchus*—soybean interactions compared with that available on root-knot and cyst nematodes. That situation remains largely unchanged, although at least 11 *Pratylenchus* spp. parasitize soybeans (446). In one survey, 41% of soybean fields in Minnesota were infested with *Pratylenchus* spp. (186). *P. brachyurus* (Godfrey) Filipjev and Schuurmans Stekhoven is, if not the species most damaging to soybeans, the one about which most information is available. This species, however, is limited in distribution to the southern and eastern states (369). Soybean is a host of *P. vulnus* Allen and Jensen in the greenhouse, but no field association has been reported (8,446). The remaining nine species are distributed in the north central United States and have been associated with soybeans in the field: *P. agilis* Thorne and Malek; *P. alleni* Ferris; *P. coffeae* (Zimmermann) Filipjev and Schuurmans Stekhoven; *P. crenatus* Loof; *P. hexincisus* Taylor and Jenkins; *P. neglectus* (Rensch) Filipjev and Schuurmans Stekhoven; *P. penetrans* (Cobb) Filipjev and Schuurmans Stekhoven; *P. scribneri* Steiner; and *P. zeae* Graham. Those species for which soybean pathogenicity has been documented are *P. agilis* (408), *P. alleni* (156), *P. hexincisus* (572), *P. penetrans* (445), and *P. scribneri* (9,119,300).

In general, *Pratylenchus* spp. are migratory root endoparasites. *P. agilis* can be ectoparasitic in culture, but it is unknown whether it is ectoparasitic in the field (409). All juvenile stages and adult *Pratylenchus* are vermiform and range in length from 0.3-0.9 mm. Members of this genus are easy to recognize under low (40x) magnification, but species are difficult to separate. Compounding the problem of identification is that more than one species of *Pratylenchus* are often found in the same field (159,408). *Pratylenchus* spp. have blunt heads with strong, short stylets and bluntly rounded tails. There is no marked anterior sexual dimorphism in species that have males.

Pratylenchus spp. life cycles, male:female sex ratios, and pathogenicity are dependent on the species-host-environment combination (8,9,10,68,160,370,445). Under optimum conditions, most species' life cycles take about 30 days, so that several generations are possible in one growing season. Females deposit about 50 eggs in root cortical tissues. Embryogenesis, the first juvenile stage, and the first molt occur within the egg, which may remain in the root or be released into the soil when root tissues decay. From eggs hatch J2, which will molt three more times before they become adults. All life stages from J2 through adult are infective and able to move within roots and between roots and soil. Because of this migratory capability, densities in either soil or root samples alone are essentially meaningless; accurate population estimates require both. Pratylenchus spp. overwinter as eggs, juveniles, or adults in roots or soil.

Pratylenchus spp. may damage the host directly through destruction of root tissues, or indirectly through interaction with other pathogens or interference with nodule activity. Individual Pratylenchus enter roots directly and move and feed mostly within cortical tissue, but may penetrate the endodermis and feed in stelar tissues. P. agilis was observed to feed on epidermal cells (409). In most hosts, discoloration occurs in infected and surrounding cells, resulting in the lesions from which the nematode gets its common name. However, even sensitive hosts may not show evidence of lesion discoloration (512). Hypertrophy and hyperplasia do not occur in infected tissues. Cavities are created within the cortex when cell walls break down as a result of nematode feeding and may be inhabited by one or more individuals. As the nematode population of a root increases, cortical tissue becomes necrotic and is sloughed off. Lesions are frequently infected by secondary microorganisms which can contribute to the root-pruning effect. Nitrogen-fixing activity of soybean nodules was reduced when soybean cv.'Lee' was infected with P. penetrans (252).

Recommending methods of control for Pratylenchus spp. in soybeans presents problems. The first and most obvious is lack of data: what constitutes a Pratylenchus problem" in soybeans? Secondly, if more than one species is present, their host ranges may be complementary, which complicates rotation recommendations. For example, the Pratylenchus population in a field was dominated by P. zeae when corn was planted one year, and by P. brachyurus when soybeans were planted the next (144). Corn is more tolerant to nematode parasitism and when used as a rotation crop may allow nematode populations to build up, creating problems for soybeans grown in the following season (157,158). This situation is possible not only for corn, but for other crops used in soybean rotation in the north central states. For instance, common Pratylenchus spp. reproduce on corn, wheat, and oats (144,157,159,408,571). Pratylenchus spp. are often found infesting soybean fields concomitantly with H. glycines (144 and Niblack, unpublished data).

There are no Pratylenchus—resistant soybean varieties, although several workers have demonstrated differential responses to Pratylenchus spp., or differential nematode reproduction, in greenhouse studies and in field microplots (10,144,408,572). Cultivar tolerance to Pratylenchus parasitism could be exploited, given suitable methods for evaluation of tolerance in breeding programs (503).

SPIRAL NEMATODES

Helicotylenchus spp., spiral nematodes, are common in agricultural soils as well as in undisturbed soils throughout the north central states (154,155,158,186,368,371). Soybean is a good host for at least two species. H. pseudorobustus (Steiner) Golden and H. dihystera (Cobb) Sher (329,331,354,368,375,421). Both species are relatively large (0.6-1.0 mm long), robust, vermiform in all life stages, and parthenogenetic. H. dihystera has been observed feeding as an ecto-, endo-, and semi-endoparasite on soybean (375), as has H. pseudorobustus (501 and Niblack, unpublished data).

As noted by Norton (368), the common occurrence of *Helicotylenchus* spp. with other plant parasitic nematodes makes assessment of their importance difficult. *Helicotylenchus* spp. are generally not pathogenic on their own, but may contribute to disease development with other pathogens. No field studies have linked *H. pseudorobustus* directly to soybean damage, although it can build up to high densities in some fields (155,158). Differential host suitability among soybean cultivars has been demonstrated (368). In one greenhouse study, McGawley and Chapman (329) observed no differences in root weights of inoculated vs. control plants. *H. dihystera*, in contrast, damaged soybean roots significantly but had no effect on seed yield in a greenhouse study by Minton and Cairns (354); further, they found *H. dihystera* to be "visibly pathogenic" to Ogden soybean. In field microplots in a sandy loam soil, Ross et al. (421) observed a 16% yield decrease in *H. dihystera*—infested plots vs. controls, but only in the second year of the two-year study. Orbin (375) concluded that the nematode's preference for older tissue limited its potential as a soybean pathogen.

RING NEMATODES

Of the criconematid species, several ring nematodes are associated with soybean, but further information is available only for *Criconemella ornata* (Raski) Luc and Raski (38,443) and *Criconemoides simile* (Cobb) Chitwood (329,330,331).

Schmitt (443) observed differential susceptibility of soybean cultivars to *C. ornata*. Barker et al. (38) found *C. ornata* to be pathogenic to peanut but not to soybean in field plots and microplots. They attributed the large numbers of *C. ornata* recovered from soybean field samples to reproduction on weeds, and the nematode could be maintained on corn, a common rotation crop with soybean.

In greenhouse tests, McGawley and Chapman (330,331) found Custer soybean superior to Hood for reproduction of *C. simile* although Hood was more susceptible to damage. They also observed considerable differential susceptibility to *C. simile* among nine soybean cultivars, but saw no reduction in root weights of inoculated plants.

STUNT NEMATODES

The stunt nematodes *Tylenchorhynchus claytoni* Steiner, *T. martini* Fielding, *T. nudus* Allen, and *Quinisulcius acutus* (Allen) Siddiqi are frequently identified from soybean field soils (158,159,161,371,444,446). *T. claytoni* reduced Lee soybean yields by 21% compared with uninoculated plants in one year of a two-year field microplot study (29); however, Schmitt (unpublished data cited in 446) found no yield increases when nematicides were applied to *T. claytoni*-infested plots. Norton et al. (371) found densities of *T. nudus* to be more dependent on soil characteristics than on cropping history of a site.

DORYLAIMID NEMATODES

Dorylaimid parasites of soybean include *Xiphinema americanum* Cobb, a dagger nematode, and *Paratrichodorus minor* (Colbran) Siddiqi, a stubby-root nematode. Soybean is a good host for *X. americanum* (158,443) and is common in soybean fields (158,161,186,371,446). Schmitt (unpublished data) observed significant soybean yield losses caused by *X. americanum* in a sandy soil. *P. minor* was negatively correlated with soybean yields in a naturally infested field (439).

Other Nematodes

There are at least two other plant parasitic nematodes frequently identified from soybean field soils: *Hoplolaimus galeatus* (Cobb) Thorne, a lance nematode (70,371,446); and *Paratylenchus projectus* Jenkins, a pin nematode (158,159,161). No information is available on the *H. galeatus*-soybean association other than that cropping history had no effect on its population densities in Iowa soybeans (371).

In a crop rotation study, *P. projectus* densities increased on soybeans and leguminous forage crops, but were suppressed on corn (13). Custer soybean was a more suitable host than Hood for *P. projectus*, but no differential host suitability existed among seven other cultivars (329,331). *P. projectus* parasitism did not reduce root weights of inoculated plants in the greenhouse. Reproduction of *P. projectus* was suppressed in the presence of *H. pseudorobustus* and *C. simile*, an effect attributed to a reduction in root hairs caused by the latter two species. Under certain conditions not yet clearly identified, *P. projectus* may be pathogenic to soybean in the field (Niblack, unpublished data).

SUMMARY AND CONCLUSIONS

Plant parasitic nematode densities and pathogenicity to soybean are influenced by cropping history (158,408,439), soil type and other soil characteristics (68,159,370,371), soybean cultivars (8,10,38,144,330,331,368,443, 444,572), and the presence of other parasitic nematode species (331,439). Nonetheless, Ferris and Bernard (159) observed that nematode communities in 14 soybean sites distributed throughout Illinois and Indiana were "remarkably similar." If there are economic associations between the nematode species in these communities and soybean, they largely remain to be described. Evaluation of such associations may require consideration of more than one species at a time despite the complications this may entail. Data on single nematode species-soybean interactions are also necessary for interpretation of multiple species interactions. The differential soybean cultivar susceptibility to most of these nematodes adds at least one more variable to consider in studies. Yield losses possibly caused by most of the species discussed in this review may not be as devastating in individual soybean fields as those caused by some of the important nematode pathogens, but, by being more widespread and less obvious, may be just as expensive overall. As long as the potential for loss exists, we need to continue to acquire data with which to evaluate losses and devise control recommendations.

ANTHRACNOSE OF SOYBEANS

J. B. Sinclair

Department of Plant Pathology
University of Illinois at Urbana-Champaign
1102 S. Goodwin Avenue
Urbana, IL 61801

Anthracnose of soybeans [*Glycine max* (L.) Merr.] causes considerable damage in warm, humid production areas in the tropics and subtropics, and in temperate regions when these disease conducive conditions prevail. It was first reported in Korea in 1917. The disease now has been observed in all soybean-growing areas of the world. In the U.S., the disease occurs throughout the production areas east of the Rocky Mountains, causing the greatest losses in the southern production areas. The disease can reduce stands, seed quality, and yield by 20% or more (30,212,469). Some fundamentals of the disease and causal agents involved were reviewed by Hepperly (212) and Sinclair (469). This review focuses on material published since 1980.

CAUSAL ORGANISMS

Although the disease is caused primarily by *Colletotrichum truncatum* (Schw.) Andrus and W. D. Moore, losses to anthracnose frequently are reported either without reference to the causal species involved or with the assumption that it is *C. truncatum* alone (220,222,381,484,536,551,558). Anthracnose can involve several *Colletotrichum* spp., including *C. destructivum* O'Gara (teleomorph: *Glomerella glycines* Hori), *C. gloeosporioides* Penz. [teleomorph: *Glomerella cingulata* Ston. (Spaul. & Shrenk)], and *C. truncatum* (teleomorphic state unknown). For other mycological details of these fungi see Hepperly (212) and Sinclair (469). *C. graminicola* (Ces.) Wilson also was reported to be pathogenic in soybeans (424). All but *C. graminicola* are known to be seedborne in soybeans.

The taxonomy of *Colletotrichum* is poorly defined. The taxonomic keys of Sutton (487) and von Arx (527) commonly are used to separate species. However, some important differences exist between these keys. The species occurring in dicotyledons are recognized as *forma species* of *C. truncatum* by von Arx (527), while Sutton (487) differentiates *C. truncatum* from *C. dematium* f. sp. *truncatum* by the former's saprophytic ability, narrower conidia, and occurrence on the Leguminoseae. The designation *C. truncatum* generally is used by plant pathologists to identify the soybean pathogen. *C. gloeosporioides* has a wide host range, including soybeans, but often is not included as one of the causal agents of soybean anthracnose. The teleomorph of *C. gloeosporioides* is *G. cingulata*, and both were included in both keys. *C. destructivum* also associated with soybean anthracnose, is recognized in von Arx's but not in Sutton's key. *G. glycines* is recognized as a synonym of *G. cingulata* by von Arx (527). Although suggested earlier (516), the teleomorph of *C. destructivum* was shown to be *G. glycines* after the two keys were published (316).

HOST RANGE

Species of Colletotrichum associated with soybeans have wide host ranges. Many of the isolates of C. truncatum pathogenic on soybeans were recovered from the following plants: Abutilon theophrastii Medik. (velvetleaf), Amaranthus retroflexus L. (rough pigweed), Ambrosia artemisiifolia L. (ragweed), A. triffida L. (giant ragweed), Apocynum cannabinum Jacq. (dogbane), Cassia occidentalis L. (coffee senna), Chenopodium album L. (lambsquarters), Datura stramonium L. (jimsonweed), Glechoma hederacea L. (ground ivy), Hibiscus trionum L. (Venice mallow), Ipomoea hederacea L. Jacq. (morning glory), I. purpurea (L.) Roth (morning glory), Lupinus albus L. (white lupine), L. luteus L. (yellow lupine), Medicago sp. (alfalfa), Physalis heterophylla Nees (groundcherry), Polygonum pennsylvanicum L. (smartweed), P. persicaria L. (ladysthumb), Solanum nigrum L. (black nightshade), S carolinense L. (horsenettle), Stylosanthes capitata J. Vogel, S. guianensis Sw., S hamata Taubert, S humilis Kunth, S. macrocarpa S. F. Blake, S. scabra J. Vogel, and Xanthium pennsylvanicum Wallr. (cocklebur) (14,188,197,204,213,477,511).

C. destructivum was isolated from alfalfa (188) and a variety of weeds (204,213).

C. gloeosporioides was isolated from alfalfa (188), Sida spinosa (prickly sida) (511), velvetleaf (213) and the Stylosanthes spp. listed previously (477).

The host range of these fungi may play an important role in the epidemiology of soybean anthracnose. This was suggested by the report of C. gloeosporioides being prevalent in soybean fields growing near citrus groves in Spain (176). However, the role of alternative hosts in the epidemiology of soybean anthracnose is not known.

STUDIES IN CULTURE

The Colletotrichum fungi are not easily isolated in pure culture because of a slimy conidial matrix. For best results, infected plant material is rinsed for 30 min in running tap water, soaked for 4 min in 0.5% NaOCl or CaOCl, and rinsed in sterile distilled water. Isolation should be done from the margins of newly-formed lesions.

Infection by Colletotrichum spp. frequently is symptomless on soybean plants and seeds during the growing season, with production of stromatic bodies, acervuli, and pycnidia occurring on dead plant parts at the end of the growing season. Fruiting structures can be induced to form from freshly infected plant tissues at any time during the growing season. After washing in running tap water for 8-12 hr and soaking in 0.5% NaOCl, as described above, rinse three times in sterile distilled water and then dip pieces for 1 min into an aqueous solution of paraquat (28.1%) diluted 1:40. Plant parts then are placed on moist cellulose pads (Kimpac) or filter paper and incubated at 100% relative humidity at 25 C. After 7 days, if mycelial growth has covered tissue surfaces, spray lightly with 95% ethanol to collapse the mycelia (59). Examine the surface of plant parts under a dissecting microscope. Caution should be taken when working with paraquat - use a hood for preparing the diluted solution and for treating plant material. Read the label. Spraying paraquat or glyphosate, both desiccant herbicides, on soybean plants in the field results in greater numbers of fruiting structures of Colletotrichum than on nonsprayed plants (81,83).

The fungus grows well on a variety of media, including malt agar (MA), acidified potato-dextrose agar (PDA), potato starch agar (PSA), oatmeal agar (OMA), torula yeast agar (TYA), V-8 agar (VEA) and a variety of nonsynthetic media (67,86,222). C. truncatum grows well on MA, PDA, PSA, TYA, and VEA (213,316,469,555). Manandhar et al. (316) found that isolates of C. destructivum produced abundant acervuli on CMA, NaCl-yeast extract agar (SYA) and SYA + 10g sucrose (SYAS), and abundant perithecia of G. glycines on PDA and SYAS at 26 C. The fungi also can be cultured on autoclaved plant parts such as stems, petioles, and seeds (212).

For culture studies, inoculated plates should be incubated under 12 hr alternating light and dark using either fluorescent or ultraviolet light, for 5 to 7 days between 25-28 C (67,204,555). Conidia germination occurs between 20-32 C (6,555).

Colonies of Colletotrichum vary in color and growth habit on culture media depending upon the medium used, environmental conditions, isolate, etc., and cannot be used for species identification (86,414). Rodriguez-Marcano and Sinclair (414) showed variability in growth rate and conidia production among three isolates of C. truncatum, with one isolate being tolerant to benomyl in culture.

There is increasing evidence of variability among isolates of the fungi associated with soybean anthracnose. The extent and genetic basis of this variability is not known. Isolates from soybeans and alternative hosts have not yet been compared.

HOST-PARASITE RELATIONSHIP

Colonization of soybean stems in the field was observed to begin at the base and move upward (6). We have observed that fruiting structures of the anthracnose fungi usually appear first on the top half of plants. Since the fungi involved become established as a latent infection and symptom expression is related to age of tissue, further studies will be needed to establish the pattern of infection on maturing plants. Conidial germination, infection peg penetration, and establishment of C. truncatum and G. glycines on and in soybean leaf tissues was studied (320). Penetration of epidermal cells by infection pegs from appressoria of both fungi was common. Hyphae were observed in and between mesophyll cells at two days and in the vascular elements three days after inoculation. Veinal necrosis was evident only on leaves inoculated with C. truncatum

C. truncatum is seedborne in soybean and may survive for more than 10 years at 5 C (466). Histopathological studies show that C. truncatum alone or in mixed infections with other soybean pathogenic fungi in soybean seeds, primarily colonizes the three layers of the soybean seed coat (296,366). The fungus rarely colonizes embryo tissues but does form acervuli in seed coat tissues (414).

Disintegration of tissues in infected leaves and seeds may be due to the production of cellulolytic and pectinolytic enzymes which are produced by C. truncatum in culture and in plant tissues (85).

Physiologic aspects of the host-parasite relationship between the various fungi associated with soybean anthracnose are not known.

EPIDEMIOLOGY

It is generally accepted that, when conditions of high temperature (above 25 C) and moisture (dew, fog, humidity, or rain) prevail, the incidence of soybean anthracnose is high. In Puerto Rico, plantings in the wet season had a higher incidence of seeds infected with Colletotrichum than those which are planted later and mature in the dry season (212,214). The presence or absence of weeds had no effect on the incidence of anthracnose in soybeans in Brazil (59) or Puerto Rico (212,344). Time of artificial inoculation had less effect on disease incidence than cultivar in the percentage of soybean seeds infected with C. truncatum (39).

Seed populations of Phomopsis sp. were inversely proportional to populations of C. truncatum in Puerto Rico (214).

Girdling caused by the threecornered alfalfa hopper (*Spissistilus festinus*) did not increase the incidence of anthracnose on soybean above nongirdled ones (432). The relationship between injuries caused by other insects and the incidence of anthracnose on soybeans has not been studied.

CONTROL

Under mist chamber conditions, 413 cultivars germplasm accessions and lines of soybeans, in maturity groups 000 to X, were evaluated for their anthracnose response; Manandhar et al. (317,318) reported resistance in: Early White Eyebrow (0), Mandarin (Ottowa) (0), Mandarin (I), Waseda (II), Boone (IV), PI 95.860 (VI), and Tarheel Black (VII). In field evaluations of 1,500 accessions of maturity groups V to VII in Texas, Bowers (55) rated 43 lines as resistant with the most resistant ratings for PI's 224.273 (VII), 229.358 (VII), 416.764 (VIII), 417.061 (VIII), and 417.470 (VIII). Backman et al. (30) in Alabama and Wong et al. (555) in Malaysia noted a range of susceptibility among cultivars tested in the field.

Anthracnose of soybean seedlings increased with increased concentration of inoculum (410) and decreased with application of various calcium compounds ([CaSO $CaCO_4$, $Ca(OH)_2$] to either soil or hydroponic solutions (357). The incidence of *C truncatum* on soybean pods increased from 14-42% under field conditions to 16-75% after artificial inoculation (344). Cultural variation was reported among isolates of *C. truncatum* (86,414). These results suggest that there are many factors that can affect the evaluation of soybean germplasm for resistance. If differences in pathogenicity among isolates of *C. truncatum* are confirmed, then various germplasm lines will have to be re-evaluated for resistance.

Resistance to anthracnose of other plant parts may be inherited independent of the seedborne phase of the disease. In Brazil, Barreto (39) found that, irrespective of time of inoculation, Bragg and Parana produced seeds with lower percentage infection by *C. truncatum* than those of Santa Rose and Hardee. Hepperly et al. (214) found, among 24 lines tested in Illinois, that all but five showed little or no seed transmission of *C. truncatum* In Nepal, Manandhar et al. (319) reported soybean accessions of maturity groups 00 to VII, had increased seed infections of *C. truncatum* with increased delay in harvest.

Considerable research has been done in the southern U.S. on the use of foliar fungicides for the control of anthracnose on soybeans (13,15,212,220,222, 381,536,543,544,545,546,551,558). Recent results will be summarized here, but details on the fungicides studied, concentrations, rates, timing, cultivars used, and other information, should be obtained from original publications. Results vary between location and years. Generally, those foliar fungicides that have shown disease control and resulted in increases in seed quality and yield include: benomyl, fentin hydroxide, mancozeb, maneb, propiconazol, and thiophanate-methyl, and experimental compounds DPX 965, GX 077, TD 2126, TD 2127, TD 2179, and TPTH. Whitney (545) found that the effectiveness of various fungicides can depend upon the cultivar used in addition to environmental influences and technical differences. Control under similar conditions differed among cultivars. Almeida (15) found that in culture, benomyl and thiophanate-methyl reduced growth of *C. truncatum* more than did maneb. However, he found the reverse to be true in the field (13). Backman (26) found that control of *C. truncatum* was inversely related to chlorothalonil retention on leaf surfaces when plants were sprayed either with a petroleum or soybean oil surfactant. He found differences in deposition and adhesion among cultivars. Soybean plants sprayed with gibberellic acid under greenhouse conditions tended to have less disease than those sprayed with other growth regulators (79). The use of foliar fungicides should be used for seed crops and if market prices justify the expenses.

Diaporthe/Phomopsis COMPLEX OF SOYBEANS

J. B. Sinclair

Department of Plant Pathology
University of Illinois at Urbana-Champaign
1102 S. Goodwin Avenue
Urbana, IL 61801

The Diaporthe /Phomopsis complex of soybeans [Glycine max (L.) Merr.] causes more losses than any other soybean disease or disease complex, with the possible exception of the root rot complex. The Diaporthe/Phomopsis complex causes several diseases, the three most important being pod and stem blight, stem canker, and seed decay. Members of the complex can be part of the soybean root rot complex (193,359) and can cause seedling disease (225,278,396,429).

The complex is endemic throughout the soybean-growing areas of the U.S., with one or more of the diseases being recorded throughout the soybean-growing areas of the world. This disease complex can reduce soybean stands, seed quality, and yields by 50% or more (244,290,292,469). There are several recent reviews (244,290,469) and a conference proceedings (292) concerned with this disease complex. This review covers primarily the literature published since 1980.

CAUSAL AGENTS

The taxonomy of the Diaporthe/Phomopsis complex was reviewed recently (244). Pod and stem blight is caused primarily by Diaporthe phaseolorum (Cke. & Ell.) var. sojae Wehm. (anamorph Phomopsis sojae Leh.). Stem canker is caused primarily by Diaporthe phaseolorum (Cke. & Ell.) var. caulivora Athow and Caldwell (fertile anamorph rare). Both fungi, as well as other Phomopsis spp., may be isolated from plant parts showing symptoms of either disease. Phomopsis seed decay is caused primarily by Phomopsis longicolla Hobbs (teleomorph unknown) (244). The species of Diaporthe and Phomopsis isolated from soybean roots and seedlings were not identified (293,278,359,428). Mycological details concerned with the various fungi involved in the complex have been published (244,469). P. longicolla was recovered significantly more often from seeds than D. phaseolorum var. sojae, which was recovered more often than D. phaseolorum var. caulivora (281). In Iowa, seed infection was correlated with pod infection by the three fungi (332) while in Ohio the stem canker pathogen was isolated three times more frequently from seeds than from pods from which the seeds were taken (243). Members of the complex are highly variable, with different biotypes occurring in different geographical regions and areas. For example, it was suggested that isolates of D. phaseolorum var. caulivora from the southeastern U.S. be referred to as "southern isolates" (242). A review on stem canker appears elsewhere in this publication.

HOST RANGE

Isolates of Phomopsis spp. pathogenic to soybeans were recovered from the following plants: Amaranthus spinosus L., Leonotis nepetaefolia (L.) R. Br., and Leonurus sibiricus L. in Brazil (82); and in the U.S. from Abutilon theophrastii Medik. (velvetleaf) (213), and Canavalia ensiformis (L.) DC. (jackbean), Lotus corniculatus L. (birdsfoot-trefoil), Phaseolus aureus Roxb. (mungbean), Phaseolus coccineus L. (scarletrunner bean), Phaseolus lunatus L. (limabean), Phaseolus vulgaris L. (greenbean), Pisum sativum L. (garden pea), Vicia faba L. (broadbean), Vigna angularis (Willd.) Ohwi and Ohashi (adzuki bean), and Vigna unguiculata (L.) Walpers (cowpea) (289). Diaporthe and Phomopsis spp. from cotton (Gossypium sp.) infected soybeans and one isolate produced symptoms similar to stem canker (427). Members of the complex colonize the debris of at least 11 crops (469). The role of alternative hosts in the epidemiology of the disease complex is not known.

STUDIES IN CULTURE

The fungi of this complex grow well on potato-dextrose agar and on moist, autoclaved soybean plant parts. Phomopsis sp. did not sporulate on dead or senescent stems taken from the field at the end of the growing season, but sporulated in the next growing season after overseasoning (288). The fungi can grow on several natural and synthetic media at 15-32 C (optimum 28 C) and over a range of pH (4-7) (438). Pycnidia form under continuous light (2500 lux) or 12-hour alternating light and dark. Ferulic acid, coniferol, vanillin, and guaiacol induced sporulation in the dark (438). Development of D. phaseolorum var. sojae in culture was compared with other pyrenomycetes (259). A selective medium for D. phaseolorum var. caulivora was developed using four fungicides (392).

Infection by the Diaporthe / Phomopsis complex frequently is symptomless on soybeans and seeds during the growing season (281), with the production of stromatic bodies, perithecia, and pycnidia occurring on senescent plant parts near the end of the growing season. Fruiting structures can be induced on fresh infected plant tissues at any time during the growing season. After washing plant parts in running tap water for 8-12 hours and dipping in 0.5% NaOCl for 4 minutes, rinse three times in sterile distilled water and then dip pieces for 1 minute into an aqueous solution of paraquat (28.1%) diluted to 1:40. Plant parts then are placed on moist cellulose (Kimpac) or filter paper and incubated at 100% relative humidity and 25 C. After 7 days, if mycelial growth has covered surfaces, spray lightly with 95% ethanol to collapse the mycelia (81,83). Examine the surface of plant parts under a dissecting microscope. Caution should be taken when working with paraquat - use a hood for preparing the diluted solution and for treating plant material. Read the label. Spraying paraquat or glyphosate, both desiccant herbicides, on plants in the field results in greater numbers of fruiting structures of Diaporthe and Phomopsis than nonsprayed plants (81,83).

RESULTS OF RECENT RESEARCH

Pod and Stem Blight

Fungal colonization of soybean stems by the pod and stem blight fungus was studied. Under field conditions, colonization occurred along the stem axis without a definite pattern (6). Under greenhouse conditions, hyphae remained within 2 cm of points of inoculation until senescence, then grew terminally to 5.5 cm (235). In five soybean cultivars inoculated with Phomopsis spp., hyphae developed up to the terminal bud until flowering and then slowed; however, in the cultivar Hodgson, hyphae did not develop beyond the third node (45). The pathogen(s) is not systemic in soybeans.

Increases in soybean pod and seed infection by Phomopsis sp. has been associated with high atmospheric moisture after physiologic maturity. Seed infection rates were linearly related to water contents between 19 and 35%. No infection was observed below 19%, while at 35% seed infection responded to water potential similarly to the growth rate of Phomopsis sp. to osmotic potential on media (430). A simple model was developed for predicting infection of vegetative soybean tissue by *Phomopsis* sp., relating infection to accumulated leaf wetness (429).

Cultural practices can influence disease development. Colonization of soybean stems by *Diaporthe* was more extensive under continuous soybeans, than under a soybean-maize rotation, or under minimum tillage (6). In a U.S. study of nontarget effects of herbicides on soybean diseases, the recovery of *Phomopsis* from pod and stem tissues was less in weed-infested plots, and in plots treated with dinoseb, than other treatments (57). In Brazil, there was no effect by any herbicide or mechanical weed control treatment (59).

Control through resistance may be possible. The cultivar ISz14, developed in Hungary, had the lowest degree of infection by *D. phaseolorum* var. *sojae* when compared to three other lines (286). Ploper and Abney identify PI 417274, 416921, 417303, 417460, 404169A, 417046 and 416946 as having resistance to *Diaporthe*/*Phomopsis* seed infection (399). Of 1,000 breeding lines and 35 cultivars tested in the U.S., 11 of 21 resistant lines had a common lineage with PI 227687 and 229358. Two other resistant lines were descendants of the cultivar Santa from Brazil (43).

Bacillus subtilis inhibited mycelial growth and stroma formation of *Phomopsis* sp. in culture, and as a soybean seed treatment reduced stem infection in the field and pod infection in a growth chamber (110). However, the bacterium was pathogenic to soybeans.

Stem Canker

Stem canker has become a serious disease of soybeans in the southern soybean-growing areas in recent years. For example, as many as 81% of the fields in Florida had the disease in 1983 (397). Several studies have concluded that the biotype of the fungus causing stem canker is different from that in the northern soybean-growing areas (226,242,337,356,547). The etiology, epidemiology, and control of stem canker recently was reviewed (29).

Typical symptoms of stem canker may take 50-80 days to develop after artificial inoculation (396). If infection takes place near the top of plants, tip dieback develops; if lower on the stem, typical cankers develop (243). In soybean plants inoculated near ground level with *D. phaseolorum* var. *caulivora*. the fungus grew 19-27% as far as the lesion that developed; when inoculated near the top of the plant, the fungus grew 21-64% as far as the lesion (269). The pathogen(s) is not systemic in soybeans.

In an interaction study with the threecornered alfalfa hopper, soybean stems with and without basal girdles were inoculated at internodes two and ten with *D. phaseolorum* var. *caulivora*. Stem canker lengths were significantly greater on girdled stems than on nongirdled stems, and greater at upper than lower sites regardless of girdle status (433).

In an interaction study with a fungus, results suggested that *Phytophthora megasperma* f. sp. *glycines*-suppressive soil was also suppressive to *D. phaseolorum* var. *sojae* and that there was no synergism between the two pathogens, and that the latter can be a virulent pathogen in *Phytophthora* problem soils (225).

Control of stem canker may come through conventional tillage, use of fungicide sprays, and/or resistant cultivars. In Georgia, it was found that stem canker was significantly more severe in reduced tillage plots than in conventional tillage plots (25,422). Also, the disease was more severe in soybean-wheat double cropping than in a soybean cropping sequence (422).

A number of cultivars and breeding lines have been reported resistant to the southern stem canker pathogen, including: A78-2270016 (226), A79-331022 (226), Braxton (44,526), Coker 368 (44), Cumberland (107), D67-5677-1 (44,526), Hawkeye (226), Hawkeye 63 (226), Midwest (224,226), Pride B216 (226), Terra Veg 606 (44),

Tracy M (107,526), Williams 82 (107), Yield King 563 (44), and 15 unlisted cultivars (270). However, because of the potential biotypes of the pathogen, the effect of environment, tillage practices, and other interactions on disease development, a cultivar or breeding line found resistant in one locality may not be resistant in another (107,356).

Fungicide sprays of anilazene (380), mancozeb (380), or thiabendazole (393) were reported to control stem canker. Yields were significantly higher when the first two were applied before rather than after inoculation. The level of infection by D. phaseolorum var. caulivora decreased with increased number of sprays from three to eleven using thiabendazole.

Phomopsis Seed Decay

The factors affecting Phomopsis seed decay were reviewed recently (513). In addition to reducing seed quality, members of the complex, when seedborne, can reduce stands (143,177,183,281,293,339,383). Of nine fungal genera isolated from soybean seeds, only Phomopsis spp. was correlated with reduction of seedling emergence in the laboratory and the field (339). Correlations between D phaseolorum var. sojae and greenhouse and field emergence were highly significant (293). Seeds from 2- and 4-week delayed harvest showed a 14 and 37% reduction in field stand, respectively (383). Transmission of Phomopsis spp. from inoculated soybean seeds to seedlings was detected in the field, but transmission from dead seeds to seedlings was not (177). Seeds inoculated with either of the three fungi of the complex, or from highly infected seeds, germinated poorly in nonsterile soil (281). In growth chamber studies, Phomopsis spp. reduced emergence and stand primarily in dry soil (183).

Studies on the colonization of soybean seeds by Phomopsis spp. showed that the fungus can enter through the micropyle and hilar region or through epidermal pores or cracks in seed coat (473,526). It can invade the embryo, endosperm and seed coat tissues causing tissue disintegration (473). Thus, it may be systemic in seedlings developing from infected seeds. In a histopathologic examination of mixed infections of Phomopsis spp. with either Cercospora kikuchii (T. Matsu & Tomoyasu) Gardner or Colletotrichum truncatum (Schw.) Andrus and W. D. Moore, there was an additive effect on deterioration of seed coat tissues when both Phomopsis sp. and C. truncatum were present, but not when C. kikuchii was present (296,473).

In addition to the immediate effects of moisture and temperature, other factors may affect the occurrence of the three fungi in soybean seeds.

The occurrence of the fungi isolated most often from soybean seeds differed between the northern and southern regions of Illinois in 1978-81, with Phomopsis spp. being higher in the northern areas where rainfall was higher than in the southern areas (265). In Illinois, variation in germination correlated with the incidence of Phomopsis sp. in seeds, while in Puerto Rico, germination was correlated with the incidence of Phomopsis sp. plus C. kikuchii and Fusarium sp. (218). When soybean plants were inoculated in the field with Phomopsis sp. and Cercospora sojina Hara., alone or in combination, the two acted independently in reducing yield and seed germination, and had an additive effect (50).

In two susceptible cultivars, the percentage of infected seeds was higher in samples from the lower third of the plant, whereas in a resistant genotype, pod position did not affect seed damage (216). The stage of pod senescence was found to be critical for rapid seed colonization (217). Low levels of seed infection were found in green pods, whereas high levels were found in yellow and brown pods. Increasing the period of pod incubation increased the rate of seed infection regardless of pod maturity. Also, pubescent cultivars developed more pod lesions than sparsely pubescent ones (217).

The frequency of *Phomopsis* spp. on soybean seeds grown in a stand of common cocklebur (*Xanthium pennsylvanicum* Wallr.) was 20, 38, and 20% after 0, 8, and 16 weeks of competition, respectively, suggesting that there was a limited weed competition effect (279). In a study of the effect of 18 preplant incorporated herbicides or herbicide combinations, on Wells soybean seed quality, a combination of chloramben, metribuzin, and pendimethalin significantly increased the recovery of *Phomopsis* spp., while a combination of fluchloralin and metribuzin gave significantly lower levels (56,58). Seeds harvested from field grown plants inoculated with *Phomopsis* spp. had a lower recovery of the fungus from threshed seeds than from seeds aseptically removed from pods, and the rate of transmission was underestimated when the seeds from both lots were disinfected (566).

The level of *Diaporthe* /*Phomopsis* in seeds was decreased 10% when soybeans followed maize (*Zea mays* L.) in rotation, decreased 17% by late sowing, decreased 9% by high potassium with early sowing, and decreased 1% with high potassium and late sowing (256). Seeds from nonweeded control plots had a significantly higher incidence of *Phomopsis* spp. than seeds from all herbicide-treated plots under conventional and reduced tillage, but not from no-till plots (56). The application of potassium decreased infection by *Phomopsis* spp. more in seeds from the lower than the upper nodes (257). In Florida, seed yields decreased significantly with increased proportions of seeds with *P. sojae* in the planting seed mixture (262).

From a study involving 54 cultivar/plant date combinations in Kentucky, it was found that seed infection by *Phomopsis* spp. decreased, and standard germination and seed vigor increased, with later dates of harvest maturity (504). Late-maturing cultivars Kent and York produced the highest quality seeds. Delaying the planting of early- and mid-season cultivars, such as Beeson, Cutler, and Williams, had the same effect. However, the low incidence of seed infection by *Phomopsis* spp. in late-maturing cultivars was found to be due to escape from infection and disease development rather than inherent resistance (549).

The epidemiology of Phomopsis seed decay continues to be studied and used, in part, for predictive systems for scheduling foliar fungicides. Nine of 26 states have a point system for predictive use. Twelve recommend foliar fungicides on grain production and seven for seed production (483). In Kentucky, a model was developed to predict infection by *Phomopsis* spp., with results suggesting that infestation of seeds by the pathogen depended more on moisture than on temperature (505). In Iowa, detached pods at the R6 growth stage were used to predict seed infection potentials by members of the *Diaporthe* / *Phomopsis* complex and the application of benomyl for controlling seed infection (335). In Kentucky, measuring pod infection at R6 growth stage was used in a point system to predict seed infection by *Phomopsis* spp. (507).

Several methods have been studied for control of Phomopsis seed decay. Seed treatment, using a mixture of benomyl plus thiram fungicides, increased seedling emergence and stand in the laboratory and the field (364). Treatment with captan plus thiram increased germination in the laboratory by 18% and field emergence by 6% (522). However, another report listed no effect by seed treatment with fungicides (209). Treating soybean seeds of various cultivars with heated maize, palm, soybean, or sunflower oils decreased the recovery of *Phomopsis* spp. with a concomitant increase in germination (405,570). This method can be used for small seed lots of breeding line, in germplasm collections, and in research requiring reduced seed infection.

The use of foliar fungicides to control seedborne fungi, particularly the *Diaporthe* /*Phomopsis* complex, is widespread throughout the soybean growing areas of the U.S. (290,292,469). Recent studies show that enhanced seed germination but little yield response was obtained with a single application of benomyl at the R6 growth stage (506). In a 4-year study, it was found that the application of benomyl by various methods consistently increased seed germination but not grain yields in Ohio (258). In Georgia, it was found that it was necessary to harvest promptly at maturity to obtain high quality seeds regardless of fungicide used (522).

There are a number of reports on possible resistance and tolerance to Phomopsis seed decay (290,292). Recent studies indicated that PI 205912 possessed some resistance to *Phomopsis* sp. and PI 181550 was resistant (526). Cultivars that appeared to have resistance were Hark (217), Kent (504), 0x303 (507), Williams (217), and York (504). PI 417479 (cv. Yougetsu) was found to be highly resistant to Phomopsis seed decay (352,353). Ploper and Abney identified PI 417274, 416921, 417303, 417460, 404169A, 417046, and 416946 as having resistance to *Diaporthe*/ *Phomopsis* seed infection. Pubescent lines were thought to be more resistant than sparsely pubescent lines (217), and isolines with indeterminate and semi-determinate growth habit had less seed infection than determinate types (514). The low incidence of Phomopsis seed decay in late-maturing cultivars is not due to resistance but a tendency to escape infection (549).

Microscopic and ultramicroscopic studies suggested that the micropyle and other natural openings of the seed coat are important in the entry of seedborne fungi into seeds; when these are plugged or blocked, the seeds are resistant (526). It was suggested that in breeding for resistance, consideration should be given to whether soybean lines are either more susceptible to seed-invading fungi which prevent infection by *Phomopsis* spp. or are genetically resistant to the fungi (405).

USE AND MANAGEMENT OF RESISTANCE FOR CONTROL OF BROWN STEM ROT OF SOYBEANS

H. Tachibana

Research Plant Pathologist, Agricultural Research Service
U.S. Department of Agriculture and Adjunct Professor
Department of Plant Pathology, Weed, and Weed Sciences
Iowa State University
Ames, IA 50011

Brown stem rot (BSR) of soybean, *Glycine max* (L.) Merr., is caused by the soil borne fungal pathogen *Phialophora gregata* (Allington & Chamberlain) W. Gams. The disease is widespread, particularly in Iowa, Illinois, Indiana, Minnesota, Missouri, and Wisconsin. In diseased plants browning occurs within vascular tissue and spreads to the pith to produce the characteristic symptom of the stem. Grain yield loss as high as 44% has been reported (137). Crop rotation has been recommended for BSR control since the disease was reported in 1948 (12). Forty years ago, soybeans were one of many crops grown in the Midwest. Today, the region is primarily a corn-soybean rotation agriculture. Although rotation is practiced, the corn-soybean rotation has been inadequate for BSR control. The disease was found in 95% of fields surveyed in northern Iowa in 1977 where BSR was previously considered unimportant (494). Use of resistant cultivars became the most practical and efficient method for BSR control.

RESISTANCE

Finding of Resistance

Chamberlain and Bernard (89) first reported on resistance to BSR in 1968, although they had observed resistance to BSR in PI 84.946-2 after the disease was reported in 1948. In the interim, they had attempted to determine the inheritance of resistance but were unsuccessful because of variability in results. Variability occurred from plant-to-plant and among seasons. However, hybrid populations were observed to average more resistant plants than did susceptible commercial cultivars. They used a combination of backcrossing, bulk, and pedigree selection to develop agronomic types with the extent of stem browning approaching that of PI 84.946-2. Sebastian and Nickell (454) reported that resistance involved a single dominant gene from one cross. However, the same gene plus another nonallelic gene that duplicates or modifies the effect of the first gene occurred in a second cross. The single dominant gene was considered adequate for BSR control.

Development of Resistance

Tachibana and Card (496) found that BSR-resistant lines developed by Chamberlain and Bernard (89) had superior yields on *P. gregata*-infested land in Iowa. Maximum yield advantages ranged from 30 to 34% with the resistant lines. Similar results were obtained with resistant lines that had been developed during the 1970's through a cooperative program between the Agricultural Research

Service, USDA, and the Iowa Agriculture and Home Economics Experiment Station at Ames. The first BSR-resistant soybean cultivar, BSR 301, was released in 1978 (497). Subsequently BSR 302, BSR 201, and BSR 101 cultivars were developed and released. In addition, the germplasms A3, A4, and A8 were released for further development by industry.

Use of Resistance for BSR Survey

When yield test reports were studied for areas in which BSR-resistant and -susceptible soybeans were grown, the incidence, severity, and economic importance of the disease became further evident. At first local and state surveys were conducted to determine incidence and severity of BSR. Economic importance was extrapolated from experimental plot yield measurements by using susceptible soybeans. Subsequently, resistant cultivars have been used to measure yield effects of the disease. In 1985, the BSR-resistant cultivars, BSR 101 and BSR 201, were the highest-yielding cultivars in northern Illinois (149). An explanation for the two cultivars being the highest yielding is that BSR was present and the resistant cultivars were favored by natural selection.

During the development of BSR-resistant soybeans, particularly in recent years, it was not uncommon for a resistant line to be higher yielding at certain locations in regional tests. The results were interpreted as improvements in both resistance and yield. From 1980 to date, I have tested BSR-resistant soybeans in both infested and uninfested fields and found that the resistant soybeans usually were the highest yielding in infested fields, whereas susceptible soybeans yielded the highest when the BSR fungus was either absent or present at low levels. An experiment conducted in 1985 to determine improvement in resistance further indicated that BSR-resistant soybeans, including BSR 101 and BSR 201, yielded significantly higher than the susceptible soybeans on infested, but not on uninfested areas (492). Thus, I conclude that the current high yield of BSR-resistant soybeans in yield tests is not because they are higher yielding than the susceptibles but because BSR has become more widespread and severe, and that favors the performance of resistant soybeans.

Disease Control by Resistance

In field tests, the incidence of disease decreases significantly in infested lands after growing resistant soybeans continuously for four years (146). After 4 yr, susceptible soybeans can be grown and their maximum yield potentials realized. Thus, if crop rotation with other crops is not currently practical, BSR-resistant soybeans can be used for rotation.

In general, a resistant cultivar is often thought to be one that does not become infected. Instead, BSR-resistant soybeans are resistant because they have less stem browning than susceptible soybeans. It is hypothesized that BSR decreases with cropping of current resistant soybeans because less inoculum is produced.

Prescribed Resistant Cultivars

The concept of prescribed resistant cultivars (PRC) was first introduced with BSR 301 (491), and was the basis for the release of BSR 301. At the time of its release, BSR 301 yielded more than susceptible cultivars on infested, but not on uninfested, fields. BSR-resistant cultivars were not the highest yielding in regional tests. In the absence of superior yields in regional tests, new lines are not usually released. Prior to BSR 301 becoming a candidate for release, other BSR-resistant lines were repeatedly rejected for this reason. After finding BSR in 95% of fields in northern Iowa in 1977, something had to be done to obtain the release of BSR-resistant lines. The PRC arguments made possible the release of the first BSR-resistant line. The rationale was that BSR resistant lines had higher yields when BSR was present but not in the absence of BSR. Thus, prior diagnosis of the presence of BSR is necessary before prescribing resistant cultivars. Plant infection level of 80% was initially reported (491) before prescribing resistant cultivars. Subsequent observations indicate that the level of infection may not be so critical.

The PRC concept not only provided a method for controlling BSR, but also made possible maximum diversification of the newly introduced germplasm from PI 84.946-2. The necessity for genetic diversity continued. After the southern corn leaf blight epidemic of 1970, which was attributed to the introduction and widespread use of T-cytoplasm for male sterility in corn, the BSR-resistant gene(s) from PI 84.946-2 not only introduce a new gene(s) for a specific purpose but also provide genetic diversity to a crop considered genetically uniform. The introduction of the BSR gene(s) provided an additional possibility for diversity because not all fields required the resistance. BSR-resistant cultivars could be prescribed only for BSR problem fields. Thus, the PRC concept and practice of prescribing resistant cultivars would preclude unrestricted production of a gene or germplasm, as happened with T-cytoplasm of corn, and limit the production of the new BSR germplasm to infested fields. It essentially creates geographical diversity and precludes a fence-to-fence cropping practice of a new germplasm introduction.

Improvement in Resistance

Although a series of BSR resistant soybeans had been developed and released, each was primarily different in maturity. BSR 101 is of maturity group (MG) I, BSR 201 of MG II, and BSR 301 and BSR 302 are of MG III. These resistant cultivars were tested under the same conditions along with other resistant lines and susceptible cultivars to compare relative resistance in Iowa in 1985. BSR results for severity were analyzed by using the "area under the disease progress curve" method (510). The study indicated improvement in resistance through breeding and selection. The most recently released cultivar, BSR 101, has the highest level of BSR-resistance. Its level of resistance in 1985 was comparable to that of PI 86.150, which was found to be more resistant than PI 84.946-2 (495). When the study was repeated in 1986, PI 86.150 was significantly more resistant that BSR 101.

Brown Stem Rot and Moisture Stress Interactions

The interaction of BSR and moisture stress was studied during the development of the resistant cultivars. The effect of moisture stress on BSR development has been observed to vary from none to highly significant. On the basis of these observations, it has been hypothesized that moisture stress interacted with BSR to affect soybean yields. Evidence of this interaction was obtained in 1985 using a movable weather shelter, thermocouples, data-loggers, and computers for continuous monitoring of moisture stress. Leaf-minus-air temperatures were used to determine moisture stress. The data in 1985 indicated essentially what had been hypothesized (Tachibana, unpublished): moisture stress interacts with BSR to affect soybean yields. The current objective is to develop soybean production models with BSR and moisture stress separately and combined. Previous reports of yield loss with BSR ranging from none to 45%, are expected to be confirmed in the study. Additionally, the research should make possible more accurate determination of soybean yields based upon fewer parameters than currently used.

PUBLIC AND PRIVATE COOPERATION FOR BSR CONTROL

On the basis of what is known about BSR and changes taking place in the industry today, BSR control will necessitate greater cooperation between public and private sectors of the soybean industry. Because our studies have shown that BSR-resistant soybeans have high yields only in the presence of moderate levels of the disease, it would be unwise economically to promote a resistant cultivar to a grower whose field is uninfested; in such fields, a susceptible cultivar with higher yield potential should be promoted. Until BSR resistance is incorporated into highest-yielding varieties BSR resistant soybeans need to be prescribed per the PRC concept and high-yielding susceptible soybeans emphasized in the absence of economically significant levels of BSR. This will optimize soybean production efficiency.

The development of soybean varieties has shifted from the public to the private sector, with the public sector now emphasizing germplasm improvement and basic research. The question arises whether the private sector will also undertake disease control in a commodity of national concern with problems such as BSR and its resistance. Although every company developing new soybean varieties has a plant breeder or two, only a few companies have a plant pathologist or an individual working on developing BSR resistant soybeans. Through cooperation of the public and private sectors, BSR can be controlled with current resistance and information.

SUMMARY

Rotation, including corn-soybean, has not been successful in the control of BSR, which has continued to spread and become of greater economic importance. Although BSR resistance was known and available for many years, it was not until recently that resistant cultivars were developed and became available for commercial production. Current high yields of BSR-resistant cultivars in yield tests are attributed to increased BSR incidence and severity. Improvement in resistance continues, but resistant cultivars alone remain inadequate for control. BSR resistant cultivars need to be prescribed. Prescribing BSR resistant cultivars will not only control the disease, but also increase soybean production efficiency. The shift of varietal development from public to private sector involves more than a shift of variety development. Disease control likewise must be considered.

CHARCOAL ROT OF SOYBEANS -- CURRENT STATUS

Thomas D. Wyllie
Department of Plant Pathology
University of Missouri

BACKGROUND INFORMATION

Macrophomina phaseolina (Tassi) Goid. is reported to cause a seedling blight, root rot, or stem rot of a least 500 species of plants (469), including many wild and cultivated species. It is found world-wide and usually under conditions of high soil temperature and low soil moisture. Infection may occur anytime during the growing season and at any stage of plant development from the seedling stage to maturity. The timing of the disease cycle depends somewhat upon the suscept attacked and where it is grown: in the tropics, sub-tropics, or temperate regions of the world.

Charcoal rot of soybeans (Glycine max (L) Merrill) in Missouri has increased in severity annually since first observed in 1961. Currently, annual losses average ca. 5% on a statewide basis with individual growers experiencing upwards of 30-50% loss in some years and in some locations. All Missouri soils and locations examined are positive for the presence of the charcoal rot fungus. Other hosts of this fungus are corn (Zea mays)(L) and grain sorghum (Sorghum bicolor) (L) Moench. Both of these crops are commonly used in rotation with soybean in parts of the central United states. Cotton (Gossipium hirsutum (L) Tandojam) and tobacco (Nicotiana tabacum)(L) are also major economic hosts in this country. Tobacco is the 4th leading income crop in Indiana and is also grown in several of the north central states, including Missouri.

Soybean plants infected with the charcoal rot fungus exhibit symptoms such as lesions on the roots of seedling plants, darkening of the vascular tissue, and premature senescence of older plants. Seedlings show reddish brown discoloration at the soil line and above, with the discolored areas turning dark brown to black. Seedlings may die under hot, dry conditions, or, if wet, cool conditions prevail, the seedlings may survive but carry a latent infection. With the onset of hot, dry weather later in the growing season, the disease symptoms often reappear and accentuate. Plants infected at a later stage of development often show no symptoms until late in the growing season. Field symptoms encountered after mid-summer include; (1) premature senescence of soybean plants with the plants retaining their leaves rather than following normal abscission patterns, and (2) epidermal and sub-epidermal tissues of the tap and secondary roots take on a gray to black discoloration, with black microsclerotia visible when the epidermal tissue is removed. Lower parts of the stem may take on a gray to silvery appearance, which often continues several nodes up the stem. The silvery tissue is frequently dotted with small black microsclerotia, which are especially noticeable at the nodes of the plant, particularly on cultivars having gray pubescence. This silvery to gray appearance apparently is related to the increased nutrients available in the nodal tissue enabling increased production of the black microsclerotia. Cortical tissue of the roots and lower stem exhibits a grayish-black coloration when split due to the presence of microsclerotia imbedded in the tissue (3). Corn and sorghum plants exhibit similar symptoms, including graying of the vascular tissue. Microsclerotia adhere to the vascular bundles. In these latter two crops, damage is largely due to rotting of the stalk resulting in lodging of the plants.

DISEASE DEVELOPMENT

M. phaseolina has been classified as both a soil-inhabiting and a root inhabiting organism (476). The fungus survives in soil and crop debris in the form of small, black microsclerotia which are produced in root and stem tissue of its hosts and which are subsequently dispersed in soil as host tissue decays (457), or they may be dislodged during tillage operations (48,103,347,476). These microsclerotia constitute the primary source of inoculum (539). In order to cause infection of soybean plants, the microsclerotia must germinate either on the surface of, or in close proximity to, plant roots. The multi-celled microsclerotia germinate a few cells at a time when in proximity to plant roots. Germination occurs throughout the growing season as conditions become favorable. Thus, infection may occur in the spring, summer, or fall. Primary and secondary hyphal strands originating from germinated microsclerotia grow toward, and proliferate in, the rhizosphere (Wyllie, unpublished). Upon making contact with the plant surface, appressoria are formed primarily on the crest of the epidermal cells (16), followed by penetration of the cuticle and invasion of tissue within and between epidermal cells (17). Mechanical pressure exerted by the penetration peg is accompanied by pectinolytic enzymatic activity (17). Subsequent growth through tissue is predominately intercellular but intracellular penetration occurs as the cells begin to lose their integrity (17). This stage is followed rapidly by the formation of microsclerotia within moribund host cells. The whole process may occur within a three day period in young tissue but the time is extended as tissue matures.

Characteristically, initial invasion of the cortical tissue of the root is followed by vertical colonization of the vascular elements by fungal hyphae. This growth of fungal mycelium frequently causes no outward symptoms. Eventually, the fungus grows from the interior of the root and stem toward the outer periphery, producing microsclerotia in the tissue. Finally, near the end of the soybean growing season, i.e., after flowering and during pod set and pod fill, microsclerotia become visible under the epidermis of the root and crown tissue. It is usually not until this time that the grower becomes aware of a problem. The plants ripen quickly and prematurely. Normal leaf abscission is not initiated, leaves become chlorotic and senescent in appearance, and pods fail to fill to their potential. Seeds that are produced are smaller than normal. Following harvest and during the winter months, microsclerotia are first protected in organic residue and are then released into the soil environment as the residue decays.

ENVIRONMENTAL INFLUENCES

M. phaseolina is considered a high temperature pathogen, with the severity of disease increasing as soil temperatures increase and as moisture becomes limiting (469). Growth of the pathogen begins and disease symptoms appear on soybean between 28 and 35 C (469). Meyer et al. (347) observed the seedling disease of soybean caused by M. phaseolina to be greatest at 30-35 C. Bristow and Wyllie (62), reported severe seedling damage to soybeans at 30 C day and 24 C night temperatures in growth chamber studies on 14 soybean cultivars. Plant height, root volume, and root and shoot weight were affected and varied with the cultivar tested, although no true cultivar resistance was detected. These differences were not sustained in the mature plant.

Aspects of the soil environment have been examined for their role in the survival and activity of the pathogen. Gangopadhyay, Wyllie, and Teague (175) reported on the effects of soil bulk density and soil moisture content on the pathogen. A gravimetric moisture of w=0.15 supported microsclerotial numbers whereas higher (w=0.25) and lower (w=0.05) moisture levels did not. Microsclerotial germination was maintained only at w=0.15. High bulk density (Db=1.56) did not support fungal microsclerotial numbers as well as Db=1.30 or 1.15. Survival of microsclerotia was poorest when Db=1.56 and w=0.25 were combined. Wyllie, Gangopadhyay, Teague and Blanchar (560), reported on the effects of oxygen and carbon dioxide levels on the germination of microsclerotia of M. phaseolina. Germination was observed at oxygen concentrations of 16% or higher. Germination was delayed, but otherwise unaffected, as oxygen decreased from 21% to 16%. Carbon dioxide concentrations in this study (0.08 to 0.49%) were not high enough to account for the changes noticed in microsclerotial behavior.

Plant Nutrition and Soil Fertility

There is very little direct evidence pertaining to the role of plant nutrition or soil fertility on the behavior of the charcoal rot fungus. Rosenbrock amd Wyllie (415,416) reported in a study of 112 commercial soybean fields in Missouri that there seemed to be a statistical relationship between phosphorus (P) and potassium (K) levels and microsclerotial numbers. Zimmerman (569), investigated the role of P and K on microsclerotial numbers in pure culture and in naturally infested soil and compared his findings with the influence of the same factors on the growth of Fusarium oxysporum Schlecht. emend Snyder and Hansen and Rhizoctonia solani Kuhn (AG4), two ubiquitous competitive fungi. These studies revealed that M. phaseolina is capable of 9X the growth of F. oxysporum and R. solani in nutrient depleted conditions (minimal P and K) at a temperature of 32-33 C and is comparatively insensitive to changes in pH (Table 1).

These data suggest that M. phaseolina is a nutrient scavenger. This may give the fungus a competitive edge under conditions of reduced nutrient availability in naturally infested soils, especially at higher soil temperatures. When soil was amended with various concentrations of P and K and microsclerotial numbers determined, no direct linear effect of P or K was noticed. However, an increase in microsclerotial numbers was determined with P level in rhizosphere soil as compared to non-rhizosphere soil (Table 2). This suggests there was an increase in microsclerotia numbers by the germination of primary microsclerotia, followed by mycelial growth and the production of new secondary microsclerotia in the region of the rhizosphere. A statistically significant (5% level), concomitant increase in root infection was detected, corresponding to an increase in microsclerotial numbers and P concentration. It is tentatively concluded from these data and those of Kovoor (287), that M. phaseolina does have a competitive advantage over selected soil fungi: under nutrient deficient conditions; or high soil temperature. Increased acidity around hyphal strands resulting from elevated P content may reduce fungal lysis by soil bacteria. Combinations of factors may also occur. In this context, then, M phaseolina can and does effectively compete in the soil environment.

HERBICIDE EFFECTS

Filho and Dhingra (166) reported significant decreases in soil populations of M. phaseolina after soil incorporation of five herbicides. In 1984, Canaday et al (74) also observed slight decreases in numbers of microsclerotia of M. phaseolina with all herbicides tested, but the decreases were not statisically

Table 1. Sample means and standard errors of biomass accumulated as dried mycelia following 9 days of growth of *F. oxysporum*, *M. phaseolina* and *R. solani* in a Czapek-Dox broth medium deficient in phosphorus and potassium buffered with 1% calcium carbonate at 24-25 C and 32-33 C.

Fungus	Biomass Accumulated in Mg Dry Wt./Flask					
	24-25 C			32-33 C		
	Mean	S.E.	Final pH	Mean	S.E.	Final pH
F. oxysporum	56	6.45	7.0	13	5.72	7.2
M. phaseolina	67	17.71	4.7	118	14.25	6.9
R. solani	37	1.93	7.5	13	5.58	7.5

Data expressed as arithmetic mean of 4 replications

Table 2. Populations of microsclerotia of *M. phaseolina* in rhizosphere soil at 4 and 7 weeks after planting as affected by phosphorus fertilization compared to numbers of microsclerotia in soil before planting.

Fertilization (kg P/ha)	Microsclerotial population (No. ms/g soil) (1)		
	Time in Weeks		
	0	4	7
0	30.8a	49.5a	51.6a
112	30.7a	48.6a	58.1a
224	27.5a	52.9a	48.8a
Mean	29.7a(2)	50.4b	52.8b

(1) Microsclerotial populations are expressed as arithmetic means of 10 reps at time 0 and 5 reps at times 4 and 7 weeks.

(2) Means followed by a common letter are not significantly different at 5% level for an LSD mean separation test. Comparison in columns in time. Mean values compared by row.

significant. Our studies suggest that the presence of herbicides had no general effect on pathogen numbers or disease development. However, with moderate to severe herbicide stress, chloramben and 2,4-DB increased root colonization by M. phaseolina whereas alachlor decreased root colonization. Glyphosate and vernolate had no detectable effect. Although herbicides may have relatively minor effects on charcoal rot, what effects were seen were generally associated with root injury. Therefore, it is important that application of herbicides is done properly and at the recommended rates so as to avoid root injury.

PATHOGEN VARIABILITY

It may be assumed that any pathogen having the ability to infect 500 species of plants must also be variable. That is, there must be something in its genetic makeup and/or biochemical arsenal which enables it to be nondiscriminating in its selection of hosts. The work of Punithalingam (403) offers an explanation of the genomic variability present in M. phaseolina. He reported that, although conidiogenous cells and hyphal cells are initially uninucleate, the single nucleus undergoes mitosis and the daughter nuclei in turn divide mitotically so that conidia were observed to have up to 36 nuclei per conidium. Each of these nuclei apparently is identical to all of the rest that were derived from the first initial nucleus. As the conidia germinate to produce germ tubes, the nuclei stream into the developing hyphae, often dividing mitotically during migration. The most common chromosome number observed was six, although aneuploid, haploid, and diploid nuclei were observed in three of the M. phaseolina isolates studied. Although Punithaligam states that the mycelia are predominantly homokaryotic, it is possible that heterokaryosis and parasexualism as well as extensive heteroploidy might occur following hyphal fusions, which, after mitotic segregation and recombination, may account for the occurrence of cultural types or physiologic races reported by Dhingra and Sinclair (115).

Assuming that the above study is representative of other isolates of M. phaseolina, these nuclear and chromosomal patterns provide the fungus with the capability of adapting to a variety of conditions in its environment in both the saprophytic and parasitic stages of its life cycle. This "plasticity" may also explain why searches for resistant varieties have been unsuccessful.

CROP ROTATION AS A CONTROL PRACTICE

Given the high degree of variability observed in the fungus, crop rotation may not appear to be a useful control measure. On the contrary, recent evidence indicates that there is a marked effect on the reproductive potential of the fungus imposed upon it by a particular host (Table 3). Since these effects differ among hosts, it is reasoned that these differences might be exploitable in developing a "management" system.

Utilizing the information presented in Table 3, crop rotations of one and two years in all combinations of soybean, corn, grain sorghum, and cotton were made. Statistically significant (5% level) differences in numbers of microsclerotia were observed under a soybean-cotton-cotton-soybean rotation, but not when corn or grain sorghum were substituted for cotton. Three year rotations out of soybeans are necessary with corn or grain sorghum to insure a significant reduction in microsclerotia numbers. However, the data also suggest that numbers of microsclerotia do not increase using a one year rotation with corn, grain sorghum, or cotton. Therefore, once the population is reduced to an acceptable level, i.e., 15 microsclerotia per gram of soil,one year rotations of these crops with soybean should result in a fairly stable, acceptable level of infestation.

Table 3. Effect of monocropping on numbers of microsclerotia of *M.phaseolina* in soil.

Crop			Mean Numbers of
1982	1983	1984	Microsclerotia/g Soil
Soybean	Soybean	Soybean	27.2
Corn	Corn	Corn	16.1
Grain Sorghum	Grain Sorghum	Grain Sorghum	10.6
Cotton	Cotton	Cotton	7.5

Data are selected but represent significant values.

MONITORING POPULATIONS OF *Macrophomina phaseolina*

In an attempt to develop computer models that will predict changes or estimate numbers of microsclerotia of *M. phaseolina* in Missouri soils, 112 soybean growers located in six geographic regions of Missouri agreed to participate. Each region consisted of 14 to 25 commercial soybean fields. Field soils were sampled twice annually for numbers of microsclerotia. A variety of agronomic and soil factors were evaluated including soil type, soil fertility, cropping history, tillage practices, etc. Temperature and precipitation information was included, and all factors were placed into a computerized data bank and analyzed. Correlation coefficients Rho r, which measure the degree of association between variables, were calculated for all data. Further analyses, utilizing analysis of variance, lease significant difference, and multiple stepwise regression were run using the SAS (1982) statistical package. From this information, a series of mathematical models was developed incorporating the statistically significant data from all of the 112 commercial fields sampled. Predictive equations were developed from spring, spring and fall, and fall, and combined 1984 data on microsclerotial numbers and were evaluated with data collected in 1985 using multiple stepwise regression.

The predictive mathematical models are given below. For each equation, the following values were determined: y = number of microsclerotia predicted utilizing the information in the model; R^2 = percentage of the total variability explained by the model; F = significance of the relationship between y and the collection of X variables; S = standard error about the calculated derived line. The standard error is used to measure precision when evaluating the equations with additional data.

Model #1 -- Analysis of Spring, 1984 Data

$$\log y = 5.971 + 0.00034\ (K) - 0.01361\ (NM) + 0.000643\ (WM) - 0.47921\ (PREC)$$

$R^2 = 0.51$ F = 27.00 P = 0.0001 S = 0.080

where:

y = Log (microsclerotia +1)
K = Pounds potassium per acre
NM = Miles north from a fixed location (the intersection of 36° latitude and 89° longitude)
WM = Miles west from a fixed location (see above)
PREC = Amount of accumulated precipitation in inches collected from December, 1983 through February, 1984

Model #2 -- Analysis of Fall, 1984 Data

log y = 1.696 + 6.0616 (KCEC) - 0.000972 (P) - 0.0236 (PREC) - 0.00000979 (NMWM + 0.000957 (StpH)

R^2 = 0.60 F = 26.51 P = 0.0001 S = 0.063

where:

KCEC = Amount of potassium contributing to the cation exchange capacity
P = Amount of phosphorus in pounds per acre
PREC = Precipitation accumulated in inches from June through August, 1984
NMWM = Interaction of miles north X miles west from a fixed point (see above), e.g. 100X250
StpH = Interaction of percentage silt X soil acidity, e.g. 30% X 6.5

Model #3 -- Analysis of Combined 1984 Data

log y = 0.869 - 0.0000138 (NMWM) - 0.001498 (WM) - 0.00113 (P) + 0.0005421 (K) + 0.105 (pH)

R^2 = 0.44 F = 31.33 P = 0.0001 S = 0.09

where:

NMWM = Miles north X miles west from a fixed point (see above)
WM = Miles west from a fixed point (see above)
pH = Soil acidity

Model #4 -- Analysis of Fall, 1984 Data Using Spring, 1984 Microsclerotia Numbers

log y = -0.4430 + 0.2296 (pH) + 0.000225 (K) - 0.00835 (CL) - 0.000437 (NMWM) + 0.6634 log (S84 + 1)

R^2 = 0.81 F = 74.12 P = 0.0001 S = 0.0308

where:

CL = Percentage clay
S84 = Spring 1984 numbers of microsclerotia
All Others = As described above

Models 1, 2, and 3 represent multiple stepwise regressions of all statistically significant factors related to microsclerotial numbers for the data collected in the spring; fall; combined spring and fall, 1984 data, respectively. Model 4 is a similar multiple stepwise regression of the fall 1984 data when the spring 1984 microsclerotial numbers were known. These models account for 51, 60, 44, and 81%, respectively, of the observed variability in numbers of microsclerotia obtained from soil samples taken from 112 commercial soybean fields.

In 1985, numbers of microsclerotia were determined in soil samples collected from the same fields sampled in 1984. Validation of the 1984 models was then done by using 1985 data in the equations to determine the observed variability. There was a lack of fit for the second set of data for model 1. However, there was a relatively good fit for models 2 and 3 where 41 and 46%, respectively, of the observed variability was accounted for with these models. The best was obtained with model 4 where 85% of the observed variability was explained.

These studies suggest that by knowing something about a growers' soil type and soil fertility; the location of the farm in the state, and field history, the dynamics of microsclerotia populations can be monitored quite closely. Therefore a grower has the opportunity to manage this pathogen to assist him in making management decisions.

SUMMARY

M. phaseolina is a soil-borne pathogen having enormous capability to adapt to existing environmental conditions. Although not ordinarily a strong competitor, it competes well when other microflora are inhibited by environmental factors, i.e. pH, soil temperature, nutritional deficiency, etc. that simultaneously favors the charcoal rot pathogen.

Its unusual capacity to adapt to conditions to which it is challenged makes breeding for plant disease resistance difficult at best. However, changes in reproductive potential relative to its different hosts makes crop rotation a viable management concept. Furthermore, predictive models are available to assist the grower in combating this pathogen.

LATE SEASON DISEASES: SUMMARY AND DISCUSSION

J. L. Lockwood

Department of Botany and Plant Pathology
Michigan State University
East Lansing, MI 48824

Superficially, the five diseases discussed seem to have little in common other than their occurrence late in the growing season. Anthracnose is a disease of foliage, stems and pods, and generally is of little consequence in the northcentral region; pod and stem blight affects primarily seed quality and therefore plant stand and yields, in the next generation of plants; stem canker, though caused by a related pathogen, infects stems causing girdling cankers; brown stem rot is a vascular disease caused by a soilborne fungus; and charcoal rot is a root- and stem-rotting disease caused also by a soilborne pathogen.

In spite of these differences, the several pathogens involved all cause symptomless infections which remain latent during most of the growing season, manifesting themselves in disease relatively late in the season. This characteristic in part accounts for their occurrence as late-season diseases. All but one of the pathogens also are seed-borne: Colletotrichum truncatum, the Phomopsis group, and Macrophomina phaseolina. The same pathogens also have wide host ranges.

ANTHRACNOSE

Although the pathogen is widespread in the north central region of the U.S., the disease only occasionally is a problem here, whereas it can be important in the southern states (469,471). Apparently, in the north, infections usually remain latent long enough that symptoms are not manifested until after pods have filled.

PHOMOPSIS COMPLEX

Pod and Stem Blight

As Dr. Sinclair has pointed out, this disease persists as a significant problem, especially in seed production, in spite of many years of research (472). He discussed several approaches to managing the disease: fungicides, cultural practices and the use of resistance.

The use of fungicidal sprays, especially for seed production, is now quite common, and I am impressed with the forecasting methods developed at Iowa State University and the University of Kentucky to determine the necessity for spraying. Denis McGee has described the method developed at Iowa State in an earlier session (335,340). Briefly it involves determining the amount of pod colonization by the pathogen using a specific pod treatment and incubation procedure, and basing a decision on whether to apply a fungicide on the amount of colonization and weather forecasts. Dr. McGee is attempting to shorten the time required for the pod assay from a week to a matter of days by the use of monoclonal antibodies to detect the fungus in the pods. If successful, this will be an excellent application to plant pathology of the new techniques in immunology. Dr. McGee and a student, A. J. Balducchi (34), also have determined that continued high relative humidity and fairly high temperature (25 C) favor the movement of the pathogen from the pods into the seeds. Their data should allow the construction of a mathematical model which, coupled with pod colonization information, will predict the need to spray or not, depending on weather conditions and forecasts.

Dr. Sinclair has told us that many factors influence the amount of seed colonization by Phomopsis (472). For example, seed infection was reduced by preceding soybeans with corn in the rotation, delayed seeding, use of late season varieties, dry weather and avoidance of delayed harvests. In addition, certain herbicides increased seed infection whereas others decreased infection. Infected seeds produced greater stands when treated with a fungicide and when soil moisture was high.

Given the wide host range of the pathogen, it would seem that some attention should be given the relative importance of different weed hosts as inoculum sources. In fact, whether they have any role at all is really not known.

Genetic resistance to pod and stem blight has been reported, but so far has not been utilized successfully in a breeding program to develop resistant varieties (472). Possibly this has not been done because there exist no efficient laboratory or greenhouse screening methods, and instead evaluation of resistance must be done in the field, which is time-consuming and subject to variability. One wonders whether inoculation of excised pods from greenhouse-grown plants in the laboratory might provide a means of assessing resistance to pod infection. Development of resistant varieties could be complicated by the occurrence of different races, if they exist, which also is unknown. Pyndji and Sinclair (405) have raised the interesting question whether "resistance" might be due in part to a proneness of seed to be colonized by non-pathogens which exclude P. sojae.

Stem Canker

Knowledge of the epidemiology of this disease is largely lacking. The inoculum for stem canker appears to be ascospores derived from cotyledonary infections (469), but the source of cotyledonary infections is unclear. One would suspect that they are derived from infected seeds, but there appears to be no quantitative relation between the two. There apparently is no information concerning the relative importance of inoculum from infected seed, crop residues and possible alternative hosts. In Georgia (422), the southern version of the disease was more severe under no-till than conventional tillage, suggesting that crop residues harboring the pathogen may be a significant source of inoculum.

It also is not known what environmental conditions cause the latent infections in the petioles or stems to become virulent. If such were known, environmental conditions might be manipulated to maintain the latent condition, or at least the disease might be forecast and fungicides applied when there is a high probability of disease expression.

The existence of different biotypes in the north and south (422) suggests that other biotypes might also exist, which could complicate attempts to breed for resistance.

BROWN STEM ROT

Dr. Tachibana has recounted for us the history of the development of the BSR varieties in Iowa (493). A great deal of credit is due him for this accomplishment, which makes possible the successful growing of soybeans in areas where brown stem rot exists. He has told us of the difficulties and frustrations encountered in gaining acceptance of these varieties, because they yielded somewhat less than susceptible varieties in the absence of disease. Acceptance was finally achieved because of three factors: the recognition a decade ago that nearly all fields in northern Iowa had become infested with the pathogen, which required that something be done; by his forwarding the concept of 'prescribed resistant varieties,' by which the resistant varieties are recommended for use only in areas where the disease exists; and his persistence in promoting the BSR varieties. It is significant that the BSR varieties have demonstrated that it is not necessary to have immunity in order to achieve the maximum yield potential in the presence of disease. We should ponder Dr. Tachibana's concern that cooperation between public and private breeders be maintained to assure the incorporation of disease resistance into new varieties, now that the soybean seed industry has assumed major responsibility for soybean varietal development.

In addition to permitting soybeans to be grown in areas where brown stem rot exists, two other benefits have been derived from the use of the BSR varieties. Their inclusion in variety trials has provided a means of assessing the importance of brown stem rot. In some areas these varieties have out-yielded susceptible varieties, indicating that the disease was more widespread and causing greater losses than was previously recognized (493). The use of BSR varieties appears to reduce the population of the pathogen in soil. After four years of growing BSR varieties, susceptible varieties could again be grown and achieved their maximum yield potential, much as in a rotation with a non-susceptible crop (146). In this connection, it was found in Wisconsin (346) that the biomass of _Phialophora gregata_ in stems of BSR-lines was much less than that in stems of susceptible varieties. I think this is a very significant point, made all the more so by the fact that such benefits were derived from varieties that carry only partial resistance.

The development of varieties resistant to brown stem rot has been slow and painstaking, due in part to lack of a laboratory or greenhouse method of assessing resistance. In Iowa, Dr. Tachibana has used stem symptoms in the field in the development of the BSR varieties (491), whereas the Illinois group has used foliar symptoms (455). Mengistu, Grau and Gritton (346) evaluated brown stem rot-resistant lines derived using both criteria in the field in Wisconsin, and found that their performance was similar. The Iowa lines developed by stem symptoms had few foliar symptoms and the Illinois lines selected by foliar symptoms had little stem discoloration. This would suggest that foliar symptoms can be used as the screening criterion. In fact, the group at Illinois (455) has recently proposed counting the number of nodes showing leaf symptoms in greenhouse-grown plants as an objective method of assessing resistance to brown stem rot. Leaf symptoms in the greenhouse were more highly correlated with disease in the field than were the numbers of nodes showing vascular browning in the greenhouse. The advantages of being able to screen for resistance in the greenhouse include, a) a more controllable environment than occurs in the field, thus assuring that disease will occur with less variability than in the field; b) disease assessment can be done in the winter time, when field work does not compete for time; c) counting nodes having foliar symptoms is easier than splitting stems, and d) the method is non-destructive to the plants. They stress that type I isolates - those that cause defoliation in addition to vascular browning - should be used. These are more virulent than type II isolates, which cause only vascular browning. The existence of the two pathogenic types was reported by Gray in 1971 (192), and recently has been confirmed among Wisconsin isolates by Mengistu and Grau (345).

The pathogen, _P. gregata_ is a difficult one to work with. It grows slowly and therefore cultures are subject to contamination, and other fungi that inhabit the vascular system of soybeans can be easily be mistaken for it. _Acremonium_ sp., a fungus causing limited vascular browning but no foliar symptoms or reduction in plant height, was isolated by Mengistu and Grau (345) from 2-3% of soybean stems and roots. In mixed cultures of _Acremonium_ with _P. gregata_, _Acremonium_ tends to dominate. They emphasize the importance of correct identification of the pathogen and of establishing pure cultures. However, the existence of *Acremonium*, and also the more ubiquitous non-pathogenic fungus *Gliocladium roseum* (360) in soybean stems, raises the question whether they might have the capacity to exert antagonistic activity towards *P. gregata*, and thus could be exploited for biological control.

The differentiation of isolates into defoliating and non-defoliating types has been mentioned. What about races? Will the existing BSR varieties maintain their resistance or will new races of the pathogen appear that can cause damage and yield reduction in these varieties?

Nearly all of the research on brown stem rot has been related to resistance, to the neglect of the basic biology of the pathogen. The fungus has no known persistent structures, and is presumed to survive as mycelium in soybean residues. It apparently does not persist as long as some soilborne pathogens, because long rotations with corn (491) or with resistant soybean varieties (146) will allow the production of a healthy crop when followed with a susceptible variety. If it persists as mycelium in residues, it might be expected that conservation tillage practices in which soybean residues remain on the soil surface and therefore persist longer, might result in longer survival of the pathogen than when they are plowed under.

What about alternative hosts? Does *P. gregata* persist as a pathogen or weak parasite in other plants, or as a saprophyte in residues of plants other than soybean? If it does, is this important epidemiologically? Of several crop plants tested in early work (12), the pathogen caused disease only in soybean and mung bean, but extensive tests have not been made.

The disease is supposed to cause greatest losses when there is adequate moisture early and dry conditions later in the growing season (491), but in Wisconsin Dr. Grau (personal communication) found that irrigation favored the development of foliar symptoms and these symptoms were associated with the greatest losses.

The pathogen has given variable responses with respect to temperature. In Wisconsin (345), some isolates caused more severe symptoms at 20 C than at 28 C, whereas others were more severe at the higher temperature, and with still others it made no difference. Is the distribution of isolates with different temperature optima of epidemiological importance?

We are sadly lacking in knowledge of the biology of *P. gregata*. This is regrettable since such knowledge sometimes suggests management strategies that can be used to minimize disease.

CHARCOAL ROT

Dr. Wyllie has given us an excellent overview of this disease, which lacks effective means of control (559). Dr. Wyllie and his research group have focused recent research on soil populations of the pathogen, since other things being equal, disease expression should be related to populations. Several factors have been found to be associated with reduced populations in soil. These include a) rotations with three years of corn or sorghum and especially cotton (whereas two years gave inconsistent results except with cotton), b) high K and low P concentrations in soil, c) adequate precipitation in June, July and August, d) field locations north and northwest of Kansas City in Missouri, e) high soil pH, and f) previous low populations. These factors have been incorporated into models for predicting the population density in soil and one model accounted for >80% of the variability in soil populations in 1985. This information should be of great value to growers in deciding whether or not, and where, to plant soybeans, and eventually might be employed in an integrated management system for reducing populations.

This is admirable work, but there may be some limitations in its applicability to the prediction of disease. Pearson et al. (387) at Kansas State University have found that the growth of isolates of Macrophomina phaseolina obtained from soybean tissues or from soybean field soil differed physiologically from those obtained from corn tissue. Among the soybean isolates two morphological types were observed. Whether these differences are associated with host-specificity has not yet been determined. A potential for variability in M. phaseolina is suggested by differences in cultural characteristics and in virulence of different isolates (115). The basis for such variability may lie in the variable numbers of nuclei that occur in conidia or vegetative cells of M. phaseolina and the different levels of ploidy that exist (403). It should also be recognized that sometimes quite large differences in populations are not expressed in disease. For example in work in Kansas (388), soil was fumigated in the field with methyl bromide, which reduced the soil population to only 20% of that in nonfumigated soil. Naturally-infested soil also was supplemented with laboratory-grown sclerotia. Yet, neither treatment resulted in any change in disease incidence.

Varieties resistant to charcoal rot are so far not available, although variability has been seen among cultivars in disease reaction and in populations of the pathogen in asymptomatic plants. The search for resistance has been frustrated by lack of a satisfactory laboratory or greenhouse assay for screening germ plasm. The pathogen can cause a seedling rot, but the seedling reaction apparently will not predict mature plant reaction (62). Since the disease does not manifest itself beyond the seedling stage until late in the season, evaluations in the greenhouse or the field are time-consuming. Even if conditions favoring disease are met, notably high temperature and low soil moisture, disease may not occur with the precision required for selection of resistant individuals.

The difficulty in obtaining disease reliably has led to an alternative method: the measurement of fungal colonization of the root system. This method is less variable and operates under the assumption that greater colonization should result in greater disease. Work at Kansas State (388) has shown a strong inverse relationship between yield and amount of colonization in a field in which disease occurred, i.e., varieties that were colonized less suffered less yield loss. This appears to be a promising approach, but more knowledge is needed concerning the limits of its precision and its reproducibility. The measurement of root colonization is a tedious procedure requiring the grinding of root samples and plating them on a culture medium. Possibly, as is being sought in the case of Phomopsis sp., monoclonal antibodies specific to M. phaseolina could be developed which would make detection and quantification of the pathogen in soybean tissues easier and more rapid, and thus facilitate its use in a breeding program.

Dr. Wyllie and Dr. Novacky at Missouri (personal communication) are currently exploring the possibility that changes in the electrical potential of protoplast membranes from soybean root cap cells in response to metabolities of the pathogen might be a reflection of resistance.

Even if reliable and precise methods for assessing resistance to charcoal rot are developed, resistance may not be easy to find. If any of the current commercial varieties carried high resistance to charcoal rot, this probably would have been seen in field experience. It may be necessary to pyramid genes giving small increments of resistance to achieve an acceptable level of performance.

Epidemiological studies of necessity have mostly used stem colonization rather than disease expression to measure effects. In a recent study in Missouri (74), herbicides affected the amount of root colonization of soybeans by M. phaseolina. Alachlor and trifluralin decreased root colonization, whereas chloramben increased it. Several others had no effect. Thus, herbicide selection might be utilized in the management of this pathogen.

Macrophomina phaseolina persists in soil through the long-lived microsclerotia and as a parasite on other hosts, perhaps often without causing disease symptoms. The microsclerotia constitute the primary inoculum for infection, whereas the alternative hosts presumably provide a means for recycling and renewal of the pathogen. Whether *M. phaseolina* has a significant saprophytic existence in or on soil is not known. Such existence could be important in disease epidemiology if, for example, the pathogen continues to colonize infected but non-diseased soybeans or alternative hosts after their death, thereby increasing its biomass and ultimately producing microsclerotia on the roots. If such saprophytic colonization occurs, research might be directed towards restricting it, through biological or other means.

The same epidemiological consequences could also derive from competitive saprophytic colonization of plant debris of any kind, but whether *M. phaseolina* has any significant capability in this regard also is not known.

The fact that infection by *M. phaseolina* can occur early and may remain latent for a long period of time suggests that preemptive colonization of roots by avirulent strains might exclude virulent strains and induce biological control. This would seem to be a possibility worthy of evaluation. Work in Israel also suggests the possibility that biological control with antagonistic microorganisms may be of benefit (141). Seed treatment of melons and corn with *Trichoderma harzianum* reduced the incidence of charcoal rot in the field by 22 and 28%, respectively.

Given the considerable quantity of knowledge already at hand concerning charcoal rot and the prospect for additional knowledge, there would seem to be an opportunity for its management, in the future if not now, through an integrative approach. Factors to be considered could include the following: those affecting soil inoculum densities, where much information is being accumulated; environmental factors affecting disease expression; cultural practices such as herbicide selection; the use of cultivars with partial resistance (once developed); knowledge of the epidemiological importance of saprophytic survival; and the use of biological control.

LITERATURE CITED

1. Abawi, G. S., and Grogan, R. G. 1975. Source of primary inoculum and effects of temperature and moisture on infection of beans by Whetzelinia sclerotiorum. Phytopathology 65:300-309.
2. Abawi, G. S., Polach, F. J., and Molin, W. T. 1975. Infection of bean by ascospores of Whetzelinia sclerotiorum. Phytopathology 65:673-678.
3. Abney, T. S. 1976. Use of foliar fungicides on soybeans in the northern states. Proc. Soybean Seed Res. Conf. 6:6-10.
4. Abney, T. S., Balles, J. B., and Richards, T. L. 1984. Influence of growth regulators on soybean pod maturity and fungal seed infection. Abstracts World Soybean Research Conference III, Ames, Iowa. Abstract No. 102.
5. Abney, T. S., and Ploper, L. D. 1988. Seed disease: in T. D. Wyllie, and D. H. Scott eds. Soybean Diseases of the North Central Region. APS Press
6. Abney, T. S., and Richards, T. L. 1982. Fungal colonization of soybean stems. (Abstr.) Phytopathology 72:995.
7. Abney, T. S., and Luedders, V. D. 1976. Influence of temperature on deterioration of soybean seed quality. (Abstr.) Proc. Am. Phytopathol. Soc. 3:293.
8. Acosta, N. A., and Malek, R. B. 1979. Influence of temperature on population development of eight species of Pratylenchus on soybean. J. Nematol. 11:229-232.
9. Acosta, N. A., and Malek, R. B. 1981. Symptomatology and histopathology of soybean roots infected by Pratylenchus scribneri and P. alleni. J. Nematol. 13:6-12.
10. Acosta, N. A., Malek, R. B., and Edwards, D. I. 1979. Susceptibility of soybean cultivars to Pratylenchus scribneri. Journal of Agriculture of the University of Puerto Rico 63:103-110.
11. Agarwal, V. K. 1981. Seed-borne fungi and viruses of some important crops. Research bulletin 108. Pantnagar, India. Govind Ballabh Pant University of Agriculture and Technology.
12. Allington, W. B., and Chamberlain, D. W. 1948. Brown stem rot of soybean. Phytopathology 38:793-802.
13. Almeida, A. M. R. 1981. [Evaluation of the curative and preventive effect of fungicides on soybean]. Fitopatologia Brasileira 6:173-178 (in Portuguese).
14. Almeida, A. M. R., Mordardo, A., Derpsh, R., and Laffranch, S. H. 1981. [Host for soybean pathogens within plant species used as winter green manure]. Fitopatologia Brasileira 6:109-113 (in Portuguese, English summary).
15. Almeida, A. M. R., and Yamashita, J. 1977 [Evaluation of toxicity of fungicides on soybean pathogens, in vitro]. Fitopatologia Brasileira 2:211-215 (in Portuguese, English summary).

16. Ammon, V., Wyllie, T. D., and Brown, M. F. 1975. Investigation of the infection process of Macrophomina phaseolina on the surface of soybean roots using scanning electron microscopy. Mycopathologica 55:77-81.
17. Ammon, V. D., Wyllie, T. D., and Brown, M. F. 1974. An ultrastructural investigation of pathological alterations induced by Macrophomina phaseolina (Tassi) Goid. in seedlings of soybean Glycine max (L.) Merrill. Physiological Plant Pathology 4:1-4.
18. Anderson, T. R. 1986. Plant losses and yield responses to monoculture of soybean cultivars susceptible, tolerant, and resistant to Phytophthora megasperma f. sp. glycinea. Plant Disease 70:468-471.
19. Anjos, J. R. N., and Lin, M. T. 1984. Bud blight of soybeans caused by cowpea severe mosaic virus in central Brazil. Plant Disease 68:405-407.
20. Athow, K. L. 1973. Fungal diseases. Pages 459-489 in: Soybeans: Improvement, Production and Uses. B. E. Caldwell, ed. American Society of Agronomy, Madison, WI. 681 pp.
21. Athow, K. L. 1985. Phytophthora root rot of soybean. Pages 575-581 in: World Soybean Res. Conf. III: Proceedings. R. Shibles, ed. Westview Press, Boulder, CO. 1262 pp.
22. Athow, K. L., and Caldwell, R. M. 1954. A comparative study of Diaporthe stem canker and pod and stem blight of soybean. Phytopathology 44:319-325.
23. Athow, K. L., and Caldwell, R. M. 1956. The influence of seed treatment and planting rate on the emergence and yield of soybeans. Phytopathology 46:91-95.
24. Ayers, W. A., and Adams, P. B. 1979. Mycoparasitism of sclerotia of Sclerotinia and Sclerotium species by Sporidesmium sclerotivorum. Can. J. Microbiol. 25:17-23.
25. Ayinla, M., and Powell, W. M. 1984. The effect of conservation tillage and rotation with grain sorghum on soybean diseases. (Abstr.) Phytopathology 74:201.
26. Backman, P. A. 1985. Effects of spray adjuvants on deposition and retention of the fungicide chlorothalonil on soybean leaf surfaces. (Abstr.) Phytopathology 75:1175.
27. Backman, P. A., Crawford, M. A., White, J., Thurlow, D. L., and Smith, L. A. 1981. Soybean stem canker: A serious disease in Alabama. Highlights Agric. Res. 28(4):6.
28. Backman, P. A., Weaver, D. B., and Morgan-Jones, G. 1985. Soybean stem canker: An emerging disease problem. Plant Disease 69:641-647.
29. Backman, P. A., Weaver, D. B., and Morgan-Jones, G. 1985. Etiology, epidemiology, and control of stem canker. Pages 589-597 in: Proceedings of the World Soybean Research Conf. III. R. Shibles, ed. Westview Press, Boulder, CO.
30. Backman, P. A., Williams, J. C., and Crawford, M. A. 1982. Yield losses in soybean from anthracnose caused by Colletotrichum truncatum. Plant Disease 66:1032-1034.
31. Bailey, J. A. 1983. Biological perspectives of host-pathogen interactions. Pages 1-32 in: The Dynamics of Host Defence, J. A. Bailey and B. J. Deverall, eds. Academic Press, New York.
32. Baker, D., Neustadt, M. H., and Zeleny, L. 1959. Relationships between fat acidity values and types of damage in grain. Cereal Chem. 36:308-311.
33. Baker, K. F., and Cook, R. J. 1974. Biological control of plant pathogens. W. H. Freeman and Co., San Francisco, CA. 433 pp.
34. Balducchi, A. J., and McGee, D. C. 1987. Environmental factors influencing infection of soybean seeds by Phomopsis and Diaporthe species during seed maturation. Plant Disease 71:209-212.

35. Balles, J., and Abney, T. S. 1983. Reaction of near-isogenic (maturity) soybean [*Glycine max* (L.) Merr.] strains to seed-borne fungal pathogens. (Abstr.) Agron. Abstr. 75:118.
36. Banttari, E. E., and Goodwin, P. H. 1985. Detection of potato viruses S, X, and Y by enzyme-linked immunosorbent assay on nitrocellulose membranes (dot-ELISA). Plant Disease 69:202-205.
37. Barbara, D. J., and Clark, M. F. 1982. A simple indirect ELISA using $F(ab^1)_2$ fragments of immunoglobulin. J. Gen. Virol. 58:315-322.
38. Barker, K. R., Schmitt, D. P., and Campos, V. P. 1982. Response of peanut, corn, tobacco, and soybean to *Criconemella ornata*. J. Nematol. 14:576-581.
39. Barreto, M. 1980. [Effect of plant inoculation time on seed transmission of *Colletotrichum dematium* f. sp. *truncata* (Schw.) Von Arx by different soyabean (*Glycine max* (L.) Merr.) cultivars]. Ph.D. Thesis, Univ. Sao Paulo, Piracicaba, Brazil (in Portuguese). 64p.
40. Bayer, E. A., Skutelsky, E., and Wilchek, M. 1979. The avidin-biotin complex in affinity cytochemistry. Methods in Enzymology 62:308-326.
41. Bean, G. A., Schillinger, J. A., and Klaman, W. L. 1972. Occurrence of aflatoxins and aflatoxin-producing strains of *Aspergillus* spp. in soybeans. Appl. Microbiol. 2:437-439.
42. Ben-Efraim, A., Lisker, N., and Henis, Y. 1985. Spontaneous heating and the damage it causes to commercially stored soybeans in Israel. J. Stored Prod. Res. 21:179-187.
43. Berger, R. D., and Hinson, K. 1984. Resistance in soybean to *Diaporthe phaseolorum* var. *sojae*. (Abstr.) Phytopathology 74:819.
44. Berggen, G. T., Snow, J. P., Harville, B. G., and Whitam, H. K. 1985. Soybean cultivar reactions to stem canker in Louisiana. (Abstr.) Phytopathology 75:498.
45. Bernaux, P. 1981. [Progression of the mycelium of *Phomopsis sojae* Wehm. inside the stem of soybean]. Agronomie 1:783-785 (in French).
46. Beuerlein, J. E., and Schmitthenner, A. F. 1986. Ohio soybean performance trials 1985. Agronomy Dept. Series 212. Ohio Coop. Ext. Serv., Columbus OH. 8 pp.
47. Bewley, J. D., and Black, M. 1985. Seeds: Physiology of development and germination. Plenum Press, NY. 367 pp.
48. Bhattacharya, M., and Samaddar, K. R. 1976. Epidemiological studies of jute diseases. Survival of *Macrophomina phaseoli* (Maubl.) Ashby in soil. Plant and Soil 44:27-36.
49. Bhattacharya, M. K., and Ward, E. W. B. 1986. Expression of gene-specific and age-related resistance and accumulation of glyceollin in soybean leaves infected with *Phytophthora megasperma* f. sp. *glycinea*. Physiol. Plant Path. 29:105-113.
50. Bisht, V. S., and Sinclair, J. B. 1985. Effect of *Cercospora sojina* and *Phomopsis sojae* alone or in combination on seed quality and yield of soybeans. Plant Disease 69:436-439.
51. Boerma, H. R., and Hussey, R. S. 1984. Tolerance to *Heterodera glycines* in soybean. J. Nematol. 16:289-296.
52. Boerma, H. R., and Phillips, D. V. 1983. Genetic implications of the susceptibility of Kent soybean to *Cercospora sojina*. Phytopathology 73:1666-1668.
53. Boland, G. J., and Hall, R. 1986. Growthroom evaluation of soybean cultivars for resistance to *Sclerotinia sclerotiorum*. Can. J. Plant Sci. 66:559-564.

54. Bonhoff, A., Loyal, R., Ebel, J., and Grisebach, J. 1986. Race: cultivar-specific induction of enzymes related to phytoalexin biosynthesis in soybean roots following infection with _Phytophthora megasperma_ f. sp. _glycinea_. Arch. Biochem. Biophys. 246:149-154.
55. Bowers, G. R., Jr. 1984. Resistance to anthracnose. Soybean Genetics Newsletter 11:150-151.
56. Bowman, J. E., Hartman, G. L., McClary, R. D., Sinclair, J. B., Hummel, J. W., and Wax, L. M. 1986. Effects of weed control and row spacing in conventional tillage, reduced tillage, and nontillage on soybean seed quality. Plant Disease 70:673-676.
57. Bowman, J. E., and Sinclair, J. B. 1987. Effect of herbicides and oat amendment on soybean disease development and seed quality. J. Phytopathol. 117:252-259.
58. Bowman, J. E., Sinclair, J. B., and Wax, L. M. 1981. Soybean seed quality affected by preplant-incorporated herbicides. (Abstr.) Phytopathology 71:862.
59. Bowman, J. E., Sinclair, J. B., and Yorinori, J. T. 1986. Effect of herbicides on soybean disease development and seed quality in the state of Parana. Fitopatol. Bras. 11:205-216.
60. Bozzola, J. J., Yopp, J., Krishnamani, M. R. S., Richardson, J., Myers, O., and Klubeck, B. 1986. Ultrastructure of sudden death syndrome, a new disease in soybeans. Proceedings of the 44th Annual Meeting of the Electron Microscopy Society of America (ed. by G. W. Bailey), p. 286.
61. Brewer, F. L. 1981. Special assessment of the soybean cyst nematode _Heterodera glycines_ problem. USDA Joint Planning and Evaluation Paper. 41 pp.
62. Bristow, P.R. and Wyllie, T. D. 1984. Reaction of soybean cultivars to _Macrophomina phaseolina_ as seedlings in the growth chamber and field. Trans. Mo. Acad. Sci. 18:5-10.
63. Brlansky, R. H., and Derrick, K. S. 1979. Detection of seed-borne plant viruses using serologically specific electron microscopy. Phytopathology 69:96-100.
64. Brown, E. A., and Minor, H. C. 1986. Soybean resistance to Phomopsis seed decay. (Abstr.) Agron. Abstr. 78:125.
65. Bryant, G. R., Durand, D. P., and Hill, J. H. 1983. Development of a solid-phase radioimmunoassay for detection of soybean mosaic virus. Phytopathology 73:623-629.
66. Bryant, G. R., Hill, J. H., Bailey, T. B., Tachibana, H., Duranl, D. P., and Benner, H. I. 1982. Detection of soybean mosaic virus in seed by solid-phase radioimmunoassay. Plant Disease 66:693-695.
67. Bryant, R. S., and Walters, H. J. 1980. Mycelial growth and conidial production of _Colletotrichum dematium_ var. _truncata_ in culture. Phytopathology 70:565-566.
68. Burns, N. C. 1971. Soil pH effects on nematode populations associated with soybeans. J. Nematol. 3:238-245.
69. Bushnell, W. R. and Rowell, J. B. 1981. Suppressors of defense reactions: A model for roles in specificity. Phytopathology 71:1012-1014.
70. Byrd, H. W., and Delouche, J. C. 1971. Deterioration of soybean seed in storage. Proc. Assoc. Off. Seed Anal. 61:41-57.
71. Calvert, L. A., and Ghabrial, S. A. 1983. Enhancement by soybean mosaic virus of bean pod mottle virus titer in doubly infected soybean. Phytopathology 73:992-997.

72. Campo, R. J., Henning, A. A., Franca Neto, J. B., Palhano, J. B., and Lantmann, A. F. 1984. Influence of seed treatment on nodulation and nitrogen fixation of soybean. In: Seminario Nacional de Pesquisa de Soja. Campinas. Resumos, p. 14.

73. Canaday, C. H. 1983. Effects of fertilization and other factors on Phytophthora rot of soybean and on natural inoculum of Phytophthora megasperma f. sp. glycinea in soil. Ph.D. dissertation. The Ohio State University, Columbus. 79 pp.

74. Canaday, C. H., Helsel, D. G., and Wyllie, T. D. 1986. Effects of herbicide-induced stress on root colonization of soybeans by Macrophomina phaseolina. Plant Disease 70:863-866.

75. Canaday, C. H., and Schmitthenner, A. F. 1982. Isolating Phytophthora megasperma f. sp. glycinea from soil with a baiting method that minimizes Pythium contamination. Soil Biol. Biochem. 14:67-68.

76. Carris, L. M. 1986. Fungi associated with Heterodera glycines in Illinois. Ph.D. thesis. University of Illinois, Urbana, IL. 211 pp.

77. Carris, L. M., Glawe, D. A., and Edwards, D. I. 1984. A comparison of fungi associated wih Heterodera glycines cysts in two Illinois soybean fields during 1983. (Abstr.) Phytopathology 74:830-831.

78. Casale, W. L., and Hart, L. P. 1986. Influence of four herbicides on carpogenic germination and apothecium development of Sclerotinia sclerotiorum. Phytopathology 76:980-984.

79. Castro, P. R. C., Kimati, H., and Moraes, R. S. 1982. [Behaviour of Glycine max cv. Davis treated wih growth regulators on inoculation with Colletotrichum dematium f. sp. truncata Revista de Agricultura 57:183-191 (in Portuguese).

80. Caviness, C. E. 1961. Effects of skips in soybean rows. Arkansas Farm Research 10(4):12.

81. Cerkauskas, R. F., Dhingra, O. D., and Sinclair, J. B. 1983. Effect of three dessicant-type herbicides on fruiting structures of Colletotrichum truncatum and Phomopsis spp. on soybean stems. Plant Disease 67:620-622.

82. Cerkauskas, R. F., Dhingra, O. D., Sinclair, J. B., and Asmus, C. 1983. Amaranthus spinosus, Leonotis nepetaefolia, and Leonnurus sibiricus: New hosts of Phomopsis spp. in Brazil. Plant Disease 67:821-824.

83. Cerkauskas, R. F., and Sinclair, J. B. 1980. Use of paraquat to aid detection of fungi in soybean tissues. Phytopathology 70:1036-1038.

84. Cerkauskas, R. F., Verma, P. R., and McKenzie, D. L. 1986. Effects of herbicides on in vitro growth and carpogenic germination of Sclerotinia sclerotiorum. Can. J. Plant Pathol. 8:161-166.

85. Chacko, S., Khare, M. N., and Agarwal, S. C. 1978. Cellulolytic and pectinolytic enzyme production by Colletotrichum dematium f. sp. truncata causing anthracnose of soybeans. Indian Phytopathology 31:69-72.

86. Chacko, S., Khare, M. N., and Agarwal, S. C. 1978. Variation in growth and morphological characters of two isolates of Colletotrichum dematium f. sp. truncata from soybean. Indian Phytopathology 31:261-262.

87. Chah, C. C., Holmquist, C. E., Carlson, C. W., Semeniuk, G., and Hesseltine, C. W. 1972. Studies on the stimulation of chick growth by soybeans molded with species of Aspergillus. (Abstr.) Poultry Sci. 51:1869.

88. Chamberlain, D. W. 1951. Sclerotinia stem rot of soybeans. Plant Dis. Rept. 35:490-491.

89. Chamberlain, D. W., and Bernard, R. L. 1968. Resistance to brown stem rot in soybeans. Crop Sci. 8:728-729.

90. Chamberlain, D. W., and Gray, L. E. 1974. Germination, seed treatment, and microorganisms in soybean seed produced in Illinois. Plant Dis. Rept. 58:50-54.

91. Chen, L-C., Durand, D. P., and Hill, J. H. 1982. Detection of soybean mosaic virus pathogenic strains by enzyme-immunosorbent assay using plates and beads as the solid phase. Phytopathology 72:1177-1181.

92. Christensen, C. M. 1967. Increase in invasion by storage fungi and in fat acidity values of commercial lots of soybeans stored at moisture contents of 13.0-14.0%. Phytopathology 57:622-624.

93. Christensen, C. M. 1973. Loss of viability in storage: Microflora. Seed Sci. Technol. 1:547-562.

94. Christensen, C. M., and Dorworth, C. E. 1966. Influence of moisture content, temperature, and time on invasion of soybeans by storage fungi. Phytopathology 56:412-418.

95. Christensen, C. M., Meronuck, R. A., Steele, J. A., and Behrens, J. C. 1973. Some morphological and chemical characteristics of binburned and fireburned soybeans. Trans. Amer. Soc. Agric. Eng. 16:899-901.

96. Christensen, C. M., and Sauer, D. B. 1982. Microflora. Pages 219-240 in: Storage of Cereal Grains and Their Products. 3rd ed. C. M. Christensen, ed. Amer. Assoc. Cereal Chem., St. Paul, MN. 544 pp.

97. Clark, B. E. 1954. Factors affecting the germination of sweet corn in low-temperature laboratory tests. N.Y. Agr. Exp. Sta. Bul. No. 769.

98. Clark, B. E., McDonald, M. B. Jr., and Joo, K. (eds). 1982. Seed Vigor Testing Handbook. Assoc. Off. Seed Analysts. Madison, WI. 173 pp.

99. Classen, D., and Ward, E. W. B. 1985. Temperature-induced susceptibility of soybeans to Phytophthora megasperma f. sp. glycinea. Physiol. Plant Path. 26:289-296.

100. Cline, M. N., and Jacobsen, B. J. 1983. Methods for evaluating soybean cultivars for resistance to Sclerotinia sclerotiorum. 1983. Plant Disease 67:784-786.

101. Coffey, M. D., and Bower, L. A. 1984. In vitro variability among isolates of six Phytophthora species in response to metalaxyl. Phytopathology 74:502-506.

102. Cohen, S. D. 1984. Detection of mycelium and oospores of Phytophthora megasperma f. sp. glycinea by vital stains in soil. Mycologia 76:34-39.

103. Cook, G. E., Boosalis, M. G., Dunkle, L. E., and Odvody, C. N. 1973. Survival of Macrophomina phaseolina in corn and sorghum stalk residue. Plant Dis. Rept. 57:873-875.

104. Cooper, E. S. 1977. Growth of the legume seedling. Adv. Agron. 29:119-139.

105. Cooper, R. L. 1971. Influence of soybean production practices on lodging and seed yield in highly productive environments. Agron. J. 63:490-492.

106. Cordonnier, M. J., and Johnston, T. J. 1983. Effects of wastewater irrigation and plant and row spacing on soybean yield and development. Agron. J. 75:908-913.

107. Cosper, B. H., Weaver, D. B., and Backman, P. A. 1985. Reaction of tropical and temperate soybean cultivars to southern isolates of Diaporthe phaseolorum var. caulivora. (Abstr.) Phytopathology 74:1176.

108. Costa, J. A., Oplinger, E. S., and Pendleton, J. W. 1980. Response of soybean cultivars to planting patterns. Agron. 72:153-156.

109. Crall, J. M. 1952. A toothpick tip method of inoculation. Phytopathology 42:4-6.

110. Cubeta, M. A., Hartman, G. L., and Sinclair, J. B. 1985. Interaction between Bacillus subtilis and fungi associated with soybean seeds. Plant Disease 69:506-509.

111. Davidse, L. C., and Kruse, H. G. 1983. Resistance to acylalanine fungicides in _Phytophthora megasperma_ f. sp. _glycinea_. Pages 135-141 in: Systemische Fuingizide und antifungale Verbindungen. H. Lyr and C. Potter, eds. Akademie-Verlag, East Berlin, German Democratic Republic.
112. Davis, K. R., Darvill, A. G., and Albersheim, P. 1986. Host-pathogen interactions. XXXI. Several biotic and abiotic elicitors act synergistically in the induction of phytoalexin accumulation in soybean. Plant Mol. Biol. 6:23-32.
113. Delouche, J. C. 1973. Seed vigor in soybeans. Pages 56-72 in: Proc. Third Soybean Research Conference.
114. Delouche, J. C., and Baskin, C. C. 1973. Accelerated aging techniques for predicting the relative storability of seed lots. Seed Sci. Technol. 1:427-452.
115. Dhingra, O. D., and Sinclair, J. B. 1978. Biology and pathology of _Macrophomina phaseolina_. Universidade Federal de Viccosa, Vicsa-Minas Gerais-Brazil. Imprensa Universitaria da U.F.V. 166 pp.
116. Diaco, R., Hill, J. H., Hill, E. K., Tachibana, H., and Durand, D. P. 1985. Monoclonal antibody-based biotin-avidin ELISA for the detection of soybean mosaic virus in soybean seeds. J. Gen. Virol. 66:2089-2094.
117. Diaco, R., Lister, R. M., Hill, J. H., and Durand, D. P. 1986. Demonstration of serological relationships among isolates of barley yellow dwarf virus by using polyclonal and monoclonal antibodies. J. Gen. Virol. 67:353-362.
118. Diaco, R., Lister, R. M., Hill, J. H., and Durand, D. P. 1986. Detection of homologous and heterologous barley yellow dwarf virus isolates wih monoclonal antibodies in serologically specific electron microscopy. Phytopathology 76:225-230.
119. Dickerson, O. J. 1979. The effects of temperature on _Pratylenchus scribneri_ and _P. alleni_ populations on soybeans and tomatoes. J. Nematol. 11:23-26.
120. Dirks, V. A., Anderson, T. R., and Bolton, E. F. 1980. Effect of fertilizer and drain location on incidence of Phytophthora root rot of soybeans. Can. J. Plant Pathol. 2:179-183.
121. Doke, N. 1983. Generation of superoxide anion by potato tuber protoplasts during the hypersensitive response to hyphal wall components of _Phytophthora infestans_ and specific inhibition of the reaction by suppressors of hypersensitivity. Physiol. Plant Path. 23:359-367.
122. Dorworth, C. E., and Christensen, C. M. 1968. Influence of moisture content, temperature, and storage time upon changes in fungus flora, germinability, and fat acidity values of soybeans. Phytopathology 58:1457-1459.
123. Dueck, J., Cardwell, V. B., and Kennedy, B. W. 1972. Physiological characteristics of systemic toxemia in soybean. Phytopathology 62:964-968.
124. Duncan, D. R., and Paxton, J. D. 1981. Trifluralin enhancement of Phytophthora root rot of soybean. Plant Disease 65:435-436.
125. Duniway, J. M. 1979. Water relations of water molds. Annu. Rev. Phytopathol. 17:431-460.
126. Dunleavy, J. 1957. Variation in pathogenicity of _Diaporthe phaseolorum_ var. _sojae_ to soybean. Iowa State J. Sci. 32:105-109.
127. Dunleavy, J. M. 1982. Effects of air temperature on disease severity and peroxidase activity of soybean leaves infected by _Peronospora manschurica_. Crop Sci. 22:623-625.
128. Dunleavy, J. M. 1983. Bacterial tan spot, a new foliar disease of soybeans. Crop Sci. 23:473-476.

129. Dunleavy, J. M. 1984. Yield losses in soybeans caused by bacterial tan spot. Plant Disease 68:774-776.
130. Dunleavy, J. M. 1985. Transmission of Corynebacterium flaccumfaciens by soybean seed. (Abstr.) Phytopathology 75:1295.
131. Dunleavy, J. M. 1985. Yield losses in soybeans caused by downy mildew. Agronomy Abstr. 96.
132. Dunleavy, J. M. 1985. Spread of bacterial tan spot of soybean in the field. Plant Disease 69:1036-1039.
133. Dunleavy, J. M. 1986. Effect of temperature on systemic spread of tan spot of soybean from seed to unifoliate leaves. (Abstr.) Phytopathology 76:1079.
134. Dunleavy, J. M. 1986. Incidence of bacterial tan spot in soybean seed from cooperative regional soybean tests. Agronomy Abstr. 126.
135. Dunleavy, J. M., Keck, J. W., Gobelman-Werner, K. S., and Thompson, M. M. 1984. Prevalence of soybean downy mildew in Iowa. Plant Disease 68:778-779.
136. Dunleavy, J. M., Keck, J. W., Gobelman, K. S., and Thompson, M. M. 1983. Prevalence of Corynebacterium flaccumfaciens as incitant of bacterial tan spot of soybean in Iowa. Plant Disease 67:1277-1279.
137. Dunleavy, J. D., and Weber, C. R. 1967. Control of brown stem rot of soybeans with corn-soybean rotations. Phytopathology 57:114-117.
138. Edwards, H. H. 1981. Light and transmission electron microscope study of brown spot of soybeans. (Abstr.) Phytopathology 71:1115.
139. Edwards, D. I., and Noel, G. R. 1981. Update: Soybean cyst nematode. Illinois Res. 23:4-5.
140. Eisenback, J. D., Hirschmann, H., Sasser, J. N., and Triantaphyllou, A. C. 1981. A guide to the four most common species of root knot nematodes (Meloidogyne spp.), with a pictorial key. Raleigh, North Carolina: North Carolina State University and the United States Agency for International Development.
141. Elad, Y., Zvieli, Y., and Chet, I. 1986. Biological control of Macrophomina phaseolina (Tassi) Goid. by Trichoderma harzianum. Crop Prot. 5:288-292.
142. Ellis, M. A., Ilyas, M. B., Tenne, F. D., Sinclair, J. B., and Palm, H. L. 1974. Effect of foliar applications of benomyl on internally seedborne fungi and pod and stem blight in soybean. Plant Dis. Rept. 58:760-763.
143. Ellis, M. A., Zambrano, O., and Paschal, E. H. 1979. Effect of pod inoculation with Phomopsis sp. on seed germination of two soybean cultivars. Page 12 in: F. T. Corbin, ed., Abstracts, World Soybean Research Conf. II. Westview Press, Boulder, CO.
144. Endo, B. Y. 1967. Comparative population increase of Pratylenchus brachyurus and P. zeae in corn and in soybean varieties Lee and Peking. Phytopathology 57:118-120.
145. Epps, J. M. 1971. Recovery of soybean cyst nematode from the digestive tracts of blackbirds. J. Nematol. 3:417-419.
146. Epstein, A. H., Hatfield, J. D., and Tachibana, H. 1983. Control of brown stem rot (caused by Phialophora gregata) with concomitant increase in yield by continuous cropping of resistant soybean. (Abstr.) Phytopathology 73:816.
147. Ersek, T., Holliday, M., and Keen, N. T. 1982. Association of hypersensitive host cell death and autofluorescence with a gene for resistance to Peronospora manshurica in soybean. Phytopathology 72:628-631.
148. Erwin, D. C. 1983. Variability within and among species of Phytophthora, Pages 149-165 in: Phytophthora: Its Biology, Taxonomy, Ecology and Pathology. D. C. Erwin, S. Bartnicki-Garcia, and P. H. Tsao, eds. American Phytopathological Society, St. Paul, MN. 392 pp.

149. Esgar, R. W., Ross, G. L., Kelley, K. A., Smyth, C. A., Carmer, S. G., Graffis, D. W., and Pepper, G. E. 1985. Performance of commercial soybeans in Illinois. Ill. Coop. Ext. Serv. Circ. 1243.

150. Farmer, E. E. Effects of fungal elicitor on lignin biosynthesis in cell suspension cultures of soybean. Plant Physiol. 78:338-342.

151. Fehr, W. R., and Caviness, C. E. 1977. Stages of soybean development. Iowa Coop. Ext. Serv. Spec. Rept. 80.

152. Fehr, W. R., Caviness, C. E., Burmood, D. T., and Pennington, J. S. 1971. Stage development descriptions for soybeans, Glycine max (L.) Merrill. Crop Science 11:929-931.

153. Ferguson, M. 1985. New disease: Root rot in soybeans. South Dakota Farm and Home Research 25:17-18.

154. Ferris, V. R., and Bernard, R. L. 1958. Plant parasitic nematodes associated with soybeans in Illinois. Plant Dis. Rept. 42:798-801.

155. Ferris, V. R., and Bernard, R. L. 1961. Seasonal variations of nematode populations in soybean field soil. Plant Dis. Rept. 45:789-793.

156. Ferris, V. R., and Bernard, R. L. 1962. Injury to soybeans caused by Pratylenchus alleni. Plant Dis. Rept. 46:181-184.

157. Ferris, V. R., and Bernard, R. L. 1967. Population dynamics of nematodes in fields planted to soybeans and crops grown in rotation with soybeans. I. The genus Pratylenchus. J. Econ. Entomol. 60:405-410.

158. Ferris, V. R., and Bernard, R. L. 1971. Crop rotation effects on population densities of ectoparasitic nematodes. J. Nematol. 3:119-122.

159. Ferris, V. R., and Bernard, R. L. 1971. Effect of soil type on population densities of nematodes in soybean rotation fields. J. Nematol. 3:123-128.

160. Ferris, V. R., Ferris, J. M., and Bernard, R. L. 1967. Relative competitiveness of two species of Pratylenchus in soybeans. (Abstr.) Nematologica 13:143.

161. Ferris, V. R., Ferris, J. M., Bernard, R. L., and Probst, A. H. 1971. Community structure of plant-parasitic nematodes related to soil types in Illinois and Indiana soybean fields. J. Nematol. 3:399-408.

162. Ferriss, R. S. 1986. Effects of pre-emergence flooding temperature on soybean seedling disease. (Abstr.) Phytopathology 76:1089.

163. Ferriss, R. S., Stuckey, R. E., Gleason, M. L., and Siegel, M. R. 1987. Effects of seed quality, seed treatment, soil source and initial soil moisture on soybean seedling performance. Phytopathology 77:140-148.

164. Fett, W. F., and Jones, S. B. 1982. Role of bacterial immobilization in race-specific resistance of soybean to Pseudomonas syringae pv. glycinea. Phytopathology 72:488-492.

165. Fett, W. F., and Osman, S. F. 1981. Antibacterial activity of the soybean isoflavanoids glyceollin and coumestrol. (Abstr.) Phytopathology 71:766.

166. Filho, E. S., and Dhingra, O. D. 1980. Effect of herbicides on survival of Macrophomina phaseolina in soil. Trans. Br. Mycol. Soc. 74:61-64.

167. Filonow, A. F., and Lockwood, J. L. 1985. Evaluation of several actinomycetes and the fungus Hyphochytrium catenoides as biocontrol agents for Phytophthora root rot of soybean. Plant Disease 69:1033-1036.

168. Forster, A. C., McInnes, J. L., Skingle, D. C., and Symons, R. H. 1985. Non-radioactive hybridization probes prepared by the chemical labeling of DNA and RNA with a novel reagent, photobiotin. Nucleic Acid Res. 13:745-761.

169. Franca Neto, J. B., Henning, A. A., and Costa N. P. 1983. Effect of different levels of purple stained seeds on seed quality and yield of soybean. In: Congresso Brasileiro de Sementes 3, Campinas. Resumos ABRATES p. 28.

170. Franca Neto, J. B., Costa, N. P., Henning, A. A., Zuffo, N. C., Barreto, J. N., and Pereira, L. A. G. 1984. Effect of planting date on soybean seed quality in the state of Mato Grosso do Sul. Pesquisa em Andamento 3, EMPAER. 9p.
171. Franca Neto, J. B., and Henning, A. A. 1984. Qualidades fisiologica e sanitaria de sementes de soja. [Physiological and pathological seed quality of soybeans] Londrina, EMBRAPA - CNPSoja, Circ. Tec. 9. 39p.
172. Friedlander, A., and Navarro, S. 1972. The role of phenolic acids in the browning, spontaneous heating and deterioration of stored soybeans. Experientia 29:761-763.
173. Frosheiser, F. I. 1957. Studies on the etiology and epidemiology of Diaporthe phaseolorum var. caulivora, the cause of stem canker of soybeans. Phytopathology 47:87-94.
174. Gangopadhay, S., Wyllie, T. D., and Luedders, V. D. 1970. Charcoal rot disease of soybean transmitted by seeds. Plant Dis. Rept. 54:1088-1091.
175. Gangopadhyay, S., Wyllie, T. D., and Teague, W. R. 1982. Effect of bulk density and moisture content of soil on the survival of Macrophomina phaseolina. Plant and Soil 68:241-247.
176. Garcia-Jimenez, J., and Altaro, A. 1985. Colletotrichum gloeosporioides: A new anthracnose pathogen of soybean in Spain. Plant Disease 69:1007.
177. Garzonio, D. M., and McGee, D. C. 1983. Comparison of seeds and crop residues as sources of inoculum for pod and stem blight of soybeans. Plant Disease 67:1374-1376.
178. Gavrechenkov, Y. D., and Sinha, R. N. 1980. Keeping quality of soybeans stored under aerobic and anaerobic conditions. Can. J. Plant Sci. 60:1087-1099.
179. Gazaway, W. S., and Henderson, J. B. 1983. Stem canker disease in soybeans. Ala. Coop. Ext. Serv. Timely Infor. Sheet. 3p.
180. Geng, S., Campbell, R. N., Carter, M., and Hills, F. J. 1983. Quality-control programs for seedborne pathogens. Plant Disease 67:236-242.
181. Ghabrial, S. A., and Shepherd, R. J. 1980. A sensitive radioimmunosorbent assay for the detection of plant viruses. J. Gen. Virol. 48:311-317.
182. Gilioli, J. L., Pereira, L. A. G., Almeida, A. M. R., and Costa, N. P. 1981. Effect of sowing depth and fungicide treatment of soybean seed on emergence in soil under different humidity conditions. Fitopatologia Brasileira 6:87-92.
183. Gleason, M. L., and Ferriss, R. S. 1984. Performance of Phomopsis - infected soybean seedlots: Influence of soil water potential. (Abstr.) Phytopathology 74:795.
184. Gleason, M. L., and Ferriss, R. S. 1985. Influence of soil water potential on performance of soybean seeds infected by Phomopsis sp. Phytopathology 75:1236-1241.
185. Golden, A. M., Epps, J. M., Riggs, R. D., Duclos, L. A., Fox, J. A., and Bernard, R. L. 1970. Terminology and identity of infraspecific forms of the soybean cyst nematode (Heterodera glycines). Plant Dis. Rept. 54:544-546.
186. Good, J. M. 1973. Nematodes. pp. 527-543 in: B. E. Caldwell, ed. Soybeans: Improvement, production, uses. Madison, Wisconsin: American Society of Agronomy.
187. Gorecki, R. J., Harman, G. E., and Mattick, L. R. 1985. The volatile exudates from germinating pea seeds of different viability and vigor. Can. J. Bot. 63:1035-1039.

188. Graham, J. H., Devine, T. E., and Hanson, C. H. 1976. Occurrence and interaction of three species of Colletotrichum on alfalfa in the mid-Atlantic United States. Phytopathology 66:538-541.

189. Grau, C. R., and Bissonette, H. L. 1974. Whetzelinia stem rot on soybeans in Minnesota. Plant Dis. Rept. 58:693-695.

190. Grau, C. R., Radke, V. L., and Gillespie, F. L. 1982. Resistance of soybean cultivars to Sclerotinia sclerotiorum Plant Disease 66:506-508.

191. Grau, C. R., and Radke, V. L. 1984. Effects of cultivars and cultural practices on Sclerotinia stem rot of soybean. Plant Disease 68:56-58.

192. Gray, L. E. 1971. Variation in pathogenicity of Cephalosporium gregatum isolates. Phytopathology 61:1410-1411.

193. Gray, L. E. 1976. Phomopsis root rot of soybeans in Illinois. (Abstr.) Proceedings of the Amer. Phytopathol. Soc. 3:290.

194. Gray, L. E. 1983. Leaf dry matter loss and pod dry matter increase of Wells soybeans infected with Septoria glycines. Plant Disease 67:525-526.

195. Gray, L. E., and Pope, R. A. 1986. Influence of soil compaction on soybean stand, yield and Phytophthora root rot incidence. Agron. J. 78:189-191.

196. Groth, D. E., and Braun, E. J. 1983. Dispersal of Xanthomonas phaseoli var. sojensis in a soybean canopy. (Abstr.) Phytopathology 73:808.

197. Gudauskas, R. T., Teem, D. H., and Morgan-Jones, G. 1977. Anthracnose of Cassia occidentalis caused by Colletotrichum dematium f. sp. truncata Plant Dis. Rept. 61:464-470.

198. Guesdon, J. L., Termyck, T., and Aurameas, S. 1979. The use of avidin-biotin interaction in immunoenzymatic techniques. J. Histochem. and Cytochem. 27:1131-1139.

199. Gulya, T., and Dunleavy, J. M. 1979. Inhibition of chlorophyll synthesis by Pseudomonas glycinea Crop Sci. 19:261-264.

200. Gupta, J. P., Erwin, D. C., Eckert, J. W., and Zaki, A. I. 1985. Translocation of metalaxyl in soybean plants and control of stem rot caused by Phytophthora megasperma f. sp. glycinea. Phytopathology 75:865-869.

201. Hahn, M. G., Bonhoff, A., and Grisebach, H. 1985. Quantitative localization of the phytoalexin glyceollin I in relation to fungal hyphae in soybean roots infected with Phytophthora megasperma f. sp. glycinea. Plant Physiol. 77:591-601.

202. Halmer, P., and Bewley, J. D. 1984. A physiological perspective on seed vigour testing. Seed Sci. Technol. 12:561-575.

203. Harman, G. E. 1983. Mechanisms of seed infection and pathogenesis. Phytopathology 73:326-329.

204. Hartman, G. L., Manandhar, J. B., and Sinclair, J. B. 1986. Incidence of Colletotrichum spp. on soybeans and weeds in Illinois and pathogenicity of Colletotrichum truncatum. Plant Disease 70:780-782.

205. Heath, M. C. 1981. A generalized concept of host-parasite specificity. Phytopathology 71:1121-1123.

206. Henning, A. 1988. Seed treatment and fungal pathogens they are designed to control: in T. D. Wyllie, and D. H. Scott eds. Soybean Diseases of the North Central Region. APS Press.

207. Henning, A. A., and Franca Neto, J. B. 1980. Evaluation problems in the germination of soybean seedlots with high incidence of Phomopsis sp. Revista Brasileira de Sementes 2:9-22.

208. Henning, A. A., and Franca Neto, J. B. 1984. Effect of Phomopsis sp. on soybean seed quality in Brazil. Pages 66-67. In: Proceedings of the Conference on the Diaporthe / Phomopsis Disease Complex of soybean. Agric. Res. Serv. U.S. Dep. Agr. 89p.
209. Henning, A. A., Franca Neto, J. de B., and Almeida, A. M. R. 1980. [Effect of chemical treatment of soybean seeds with different levels of infection by Phomopsis sojae (Leh.) on emergence.] Resultados de Pesquisa de Sojae, 1979/80:87-88. Londrina, Brazil (in Portuguese).
210. Henning, A. A., Franca Neto, J. B., and Costa, N. P. 1984. Fungicides recommended for soybean seed treatment. [Fungicidas recomendados para o tratamento de sementes de soja]. Londrina, EMBRAPA-CNPSoja. Comunicado Tecnico 31. 4p.
211. Hensarling, T. P., Jacks, T. J., Lee, L. S., and Ciegler, A. 1983. Production of aflatoxins on soybeans and cottonseed meals. Mycopathologia 83:125-127.
212. Hepperly, P. R. 1985. Soybean anthracnose. Pages 547-554 in: R. Shibles, ed., Proc. World Soybean Research Conf. III. Westview Press, Boulder, CO.
213. Hepperly, P. R., Kirkpatrick, B. L., and Sinclair, J. B. 1980. Abutilon theophrasti: Wild host for three fungal parasites of soybean. Phytopathology 70:307-310.
214. Hepperly, P. R., Mignucci, J. S., Sinclair, J. B., and Mendoza, J. B. 1982. Soybean anthracnose and its seed assay in Puerto Rico. Seed Sci. Technol. 11:371-380.
215. Hepperly, P. R., and Sinclair, J. B. 1978. Quality losses in Phomopsis-infected soybean seeds. Phytopathology 68:1684-1687.
216. Hepperly, P. R., and Sinclair, J. B. 1980. Associations of plant symptoms and pod position with Phomopsis sojae seed infection and damage in soybean. Crop Sci. 20:379-381.
217. Hepperly, P. R., and Sinclair, J. B. 1980. Detached pods for studies of Phomopsis sojae pods and seed colonization. J. Agric. Univ. P.R. 64:330-337.
218. Hepperly, P. R., and Sinclair, J. B. 1981. Relationships among Cercospora kikuchii. other seed mycoflora, and germination of soybeans in Puerto Rico and Illinois. Plant Disease 65:130-132.
219. Hermans, R., and Ziegler, E. 1984. Localization of a-mannan in the hyphal wall of Phytophthora megasperma f. sp. glycinea and it's possible relevance to the host-pathogen interaction of the fungus with soybeans (Glycine max). Phytopath. Z. 109:363-366.
220. Hershman, D. W., Bachi, P. R., and Stuckey, R. E. 1986. Evaluation of foliar fungicides for disease control and yield response of soybeans, 1985. Fungicide and Nematicide Tests 41:118.
221. Hershman, D., Stuckey, R., and Bachi, P. 1984. Sudden death syndrome - soybean. Kentucky Coop. Ext. Serv., Pest News Alert, Oct. 2, 1984.
222. Hershman, D. E., Stuckey, R. E., and Bachi, P. R. 1985. Effect of foliar fungicides on brown spot, anthracnose, pod and stem blight, and yield of soybeans, 1984. Fungicide and Nematicide Tests 40:159.
223. Heydecker, W. 1969. The 'vigour' of seeds - a review. Proc. Int. Seed Test. Assoc. 34:201-219.
224. Higley, P. M., and Tachibana, M. 1982. Resistance to stem canker of soybeans. (Abstr.) Phytopathology 72:1136.
225. Higley, P. M., and Tachibana, M. 1983. A study of a possible interaction between Diaporthe phaseolorum var. caulivora and Phytophthora megasperma var. glycinea. (Abstr.) Phytopathology 73:796.

226. Higley, P. M., and Tachibana, H. 1984. Resistance of stem canker of soybean and pathogenic specialization of the causal organism, Diaporthe phaseolorum var. caulivora. (Abstr.) Phytopathology 74:1269.
227. Hildebrand, A. A. 1948. Soybean diseases in Ontario. Soybean Dig. 8: 16-17.
228. Hildebrand, A. A. 1954. Observations on the occurrence of the stem canker and pod and stem blight on mature stems of soybean. Plant Dis. Rept. 38:640-646.
229. Hildebrand, A. A. 1956. Observations on stem canker and pod and stem blight of soybeans in Ontario. Can. J. Bot. 34:577-599.
230. Hill, E. K., Hill, J. H., and Durand, D. P. 1984. Production of monoclonal antibodies to viruses in the potyvirus group: Use in radioimmunoassay. J. Gen. Virol. 65:525-532.
231. Hill, H. J., and West, S. H. 1982. Fungal penetration of soybean seed through pores. Crop Sci. 22:602-606.
232. Hill, H. J., West, S. H., and Hinson, K. 1985. Phomopsis spp. infection of soybean seeds differing in permeability. Pages 54-58 in: Proceedings of the 1984 Conference on the Diaporthe/Phomopsis Disease Complex of Soybean. M. M. Kulik, ed. U.S. Dep. Agric., Agric. Res. Serv. 89 pp.
233. Hill, J. H., Bailey, T. B., Benner, H. T., Tachibana, H., and Durand, D. P. 1987. Soybean mosaic virus: Effects of primary disease incidence on yield and seed quality. Plant Disease 71:237-239.
234. Hill, J. H., Bryant, G. R., and Durand, D. P. 1981. Detection of plant virus by using purified IgG in ELISA. J. Virol. Method. 3:27-35.
235. Hill, H. C., Horn, N. L., and Steffens, W. L. 1981. Mycelial development and control of Phomopsis sojae in artificially inoculated soybean stems. Plant Disease 65:132-134.
236. Hill, J. H. Martinson, C. A., and Russell, W. A. 1974. Seed transmission of maize dwarf mosaic and wheat streak mosaic viruses in maize and response of inbred lines. Crop Sci. 14:232-235.
237. Hine, R. B., and Wheeler, J. E. 1970. The occurrence of some previously unreported diseases in Arizona. Plant Dis. Rept. 54:179-180.
238. Hirrel, M. C. 1983. Sudden death syndrome of soybean - a disease of unknown etiology. (Abstr.) Phytopathology 73:501-502.
239. Hirrel, M. C. 1984. Influence of overhead irrigation and row width on foliar and stem diseases of soybean. (Abstr.) Phytopathology 74:628.
240. Hirrel, M. C., McDaniel, M. C., and Ashlock, L. 1985. Facts about sudden death syndrome of soybeans. University of Arkansas Coop. Ext. Serv. Letter, August 5, 1985. 7 pp.
241. Hirrell, M. C. 1986. Sudden death syndrome of soybean: New insights into its development. Proc. of the 1986 American Seed Trade Assoc., Chicago, IL. 10 pp.
242. Hobbs, T. W., and Phillips, D. B. 1985. Identification of Diaporthe and Phomopsis isolates from soybean. (Abstr.) Phytopathology 75:500.
243. Hobbs, T. W., Schmitthenner, A. F., Ellett, C. W., and Hite, R. E. 1981. Top dieback of soybean caused by Diaporthe phaseolorum var. caulivora. Plant Disease 65:618-620.
244. Hobbs, T. W., Schmitthenner, A. F., and Kuter, G. A. 1985. A new Phomopsis species from soybean. Mycologia 77:535-544.
245. Hobe, A. M. 1981. Pathogenic variability of Phytophthora megasperma f. sp. glycinea isolated from northwest Ohio soils. M.S. Thesis, The Ohio State University. 32 pp.
246. Hoffmann, K., Finn, F. M., and Kiso, Y. 1978. Avidin-biotin affinity columns. General methods for attaching biotin to peptides and proteins. J. Amer. Chem. Soc. 100:3588-3590.

247. Holliday, M. J., and Keen, N. T. 1982. The role of phytoalexins in the resistance of soybean leaves to bacteria: Effect of glyphosate on glyceollin accumulation. Phytopathology 72:1470-1474.
248. Huang, H. C. 1977. Importance of Coniothyrium minitans in survival of sclerotia of Sclerotinia sclerotiorum in wilted sunflower. Can. J. Bot. 55:289-295.
249. Hunst, P. L., and Tolin, S. A. 1982. Isolation and comparison of two strains of soybean mosaic virus. Phytopathology 72:710-713.
250. Hunst, P. L., and Tolin, S. A. 1983. Ultrastructural cytology of soybean infected with mild and severe strains of soybean mosaic virus. Phytopathology 73:615-619.
251. Hunter, J. R., and Erickson, A. E. 1952. Relation of seed germination to soil moisture tension. Agron. J. 44:107-109.
252. Hussey, R. S., and Barker, K. R. 1976. Influence of nematodes and light sources on growth and nodulation of soybean. J. Nematol. 8:48-52.
253. Inagaki, H. 1979. Race status of five Japanese populations of Heterodera glycines. Jpn. J. Nematol. 9:1-4.
254. Irwin, M. E., and Goodman, R. M. 1981. Ecology and control of soybean mosaic virus. Pages 181-220 in: K. Maramarosch and K. F. Harris, eds. Plant Diseases and Vectors: Ecology and Epidemiology.
255. Iverson, M. L. 1986. The effect of changes in interpretive lines for damaged soybeans upon the free fatty acid content of graded soybeans. Special study No. SS-3-1986, U.S. Dept. Agric., Fed. Grain Insp. Serv. 20 pp.
256. Jeffers, D. L., and Schmitthenner, A. F. 1981. Germination and disease in soybean seed affected by rotation, plant time, K fertilization and tillage. (Abstr.) Agronomy Abstr. 1981:119.
257. Jeffers, D. L., Schmitthenner, A. F., and Kroetz, M. E. 1982. Potassium fertilization effects on Phomopsis seed infection, seed quality, and yield of soybeans. Agronomy J. 74:886-890.
258. Jeffers, D. L., Schmitthenner, A. F., and Reichard, D. L. 1982. Seedborne fungi, quality, and yield of soybeans treated with benomyl fungicide by various application methods. Agronomy J. 74:589-592.
259. Jensen, J. D. 1983. The development of Diaporthe phaseolorum var. sojae in culture. Mycologia 75:1074-1091.
260. Jiminez, B., and Lockwood, J. L. 1982. Germination of oospores of Phytophthora megasperma f. sp. glycinea in the presence of soil. Phytopathology 72:727-733.
261. Johnson, B. J., and Harris, H. B. 1967. Influence of plant population on yield and other characteristics of soybeans. Agron. J. 59:447-449.
262. Johnson, S. B., and Berger, R. D. 1982. The influence of different incidences of Phomopsis infected seed on the yield of soybeans. (Abstr.) Phytopathology 72:944.
263. Jones, F. E. 1986. Development and use of a monoclonal antibody-based enzyme immunoassay for the detection of maize dwarf mosaic virus. M.S. Thesis. Iowa State University, Ames. 163 pp.
264. Jordan, E. G. 1981. Monitoring soybean foliar diseases in Illinois in 1980. (Abstr.) Phytopathology 71:884.
265. Jordan, E. G. Manandhar, J. B., Thapliyal, P. N., and Sinclair, J. B. 1986. Factors affecting soybean seed quality in Illinois. Plant Disease 70:246-248.
266. Kamicker, T. A., and Lim, S. M. 1985. Field evaluation of pathogenic variability in isolates of Septoria glycines. Plant Disease 69:744-746.

267. Keeling, B. L. 1982. A seedling test for resistance to soybean stem canker caused by Diaporthe phaseolorum var. caulivora. Phytopathology 72:807-809.
268. Keeling, B. L. 1984. Evidence for physiologic specialization in Diaporthe phaseolorum var. caulivora. J. Miss. Acad. Sci. Suppl. 29:5.
269. Keeling, B. L. 1985. Relative pathogen growth and lesion development in soybean plants inoculated with Diaporthe phaseolorum var. caulivora. (Abstr.) Phytopathology 75:1278.
270. Keeling, B. L. 1985. Soybean cultivar reactions to soybean stem canker caused by Diaporthe phaseolorum var. caulivora and pathogenic variation among isolates. Plant Disease 69:132-133.
271. Keeling, B. L. 1985. Responses of differential soybean cultivars of hypocotyl inoculation with Phytophthora megasperma f. sp. glycinea at different temperatures. Plant Disease 69:524-525.
272. Keen, N. T. 1982. Specific recognition in gene-for-gene host-parasite systems. Advances in Plant Pathol. 1:35-82.
273. Keen, N. T., Holliday, M. J., and Yoshikawa, M. 1982. Effects of glyphosate on glyphosate on glyceollin production and expression of resistance to Phytophthora megasperma f. sp. glycinea in soybean. Phytopathology 72:1467-1470.
274. Keen, N. T., and Kennedy, B. W. 1974. Hydroxyphaseollin and related isoflavanoids in the hypersensitive resistant response of soybeans against Pseudomonas glycinea. Physiol. Plant Pathol. 4:173-185.
275. Kennedy, B. W. 1964. Moisture content, mold invasion, and seed viability of stored soybeans. Phytopathology 54:771-774.
276. Kennedy, B. W. 1984. Root rot of soybeans in Minnesota: The changing race situation in Phytophthora megasperma f. sp. glycinea. Plant Disease 68:826.
277. Kennedy, B. W., and Koukkari, W. L. 1981. Measurement of soybean injury using chlorphyll assays of cotyledons. (Abstr.) Phytopathology 71:885.
278. Killebrew, J. F., and Roy, K. W. 1984. Epidemiology and mycofloral relationships in soybean seedling disease. (Abstr.) Phytopathology 74:629.
279. Kirkpatrick, B. L., Bloomberg, J. R., and Wax, L. M. 1983. The effect of competition from common cocklebur (Xanthium pennsylvanicum) on the seedborne fungi of soybean. (Abstr.) Weed Sci. Soc. Amer. 1983:58-59.
280. Kittle, D. R., and Gray, L. E. 1979. Storage and use of Phytophthora megasperma var. sojae oospores as inoculum. Phytopathology 69:821-823.
281. Kmetz, K. T., Schmitthenner, A. F., and Ellett, C. W. 1978. Soybean seed decay: Prevalence of infection and symptom expression by Phomopsis sp., Diaporthe phaseolorum var. sojae and Diaporthe phaselorum var. caulivora Phytopathology 68:836-840.
282. Koenig, R. 1978. ELISA in the study of homologous and heterologous reactions of plant viruses. J. Gen. Virol. 40:309-318.
283. Koenig, R., and Paul, H. L. 1982. Detection and differentiation of plant viruses by various ELISA procedures. Acta. Hort. 127:147-158.
284. Kohler, G., and Milstein, C. 1975. Continuous cultures of fused cells secreting antibody of predefined specificity. Nature, London 256:495-497.
285. Kohn, L. M. 1979. A monographic revision of the genus Sclerotinia. Mycotaxon 9:365-444.
286. Kovics, G. 1980. [Susceptibility of soybean varieties to Diaporthe phaseolorum var. sojae (Phomopsis sojae), the pathogen causing stem and pod blight]. Novenyvedelem 16:461-465.
287. Kovoor, A. T. A. 1954. Some factors affecting the growth of Rhizoctonia bataticola in soil. J. Madras University 24:47-52.
288. Kulik, M. M. 1981. Cessation of sporulation by Phomopsis sojae on soybean stems with the approach of autumn and its possible significance. (Abstr.) Phytopathology 71:768.

289. Kulik, M. M. 1981. Infection of legume seedlings by Phomopsis batate, P. phaseoli, and P. sojae (Abstr.) Phytopathology 71:768.
290. Kulik, M. M. 1983. The current scenario of the pod and stem blight-stem canker-seed decay complex of soybean. J. Trop. Plant Dis. 1:1-11.
291. Kulik, M. M. 1984. Symptomless infection, persistence, and production of pycnidia in host and non-host plants by Phomopsis batatatae, Phomopsis phaseoli, and Phomopsis sojae, and the taxonomic implications. Mycologia 76:274-291.
292. Kulik, M. M. (ed.) 1985. Proceedings of the Conference on the Diaporthe/ Phomopsis Disease Complex of Soybean. USDA, ARS, Washington, DC. 89p.
293. Kulik, M. M., and Schoen, J. F. 1981. Effect of seedborne Diaporthe phaseolorum var. sojae on germination, emergence and vigor of soybean seedlings. Phytopathology 71:544-547.
294. Kulik, M. M., and Yaklich, R. W. 1982. Relationship of the appearance of soybean seeds to seed-borne infection by Diaporthe phaseolorum var. sojae and other aspects of seed quality. Seed Sci. and Technol. 10:335-342.
295. Kunwar, I. K., Manandhar, J. B., and Sinclair, J. B. 1986. Histopathology of soybean seeds infected with Alternaria alternata. Phytopathology 76:543-546.
296. Kunwar, I. K., Singh, T., and Sinclair, J. B. 1985. Histopathology of mixed infections by Colletotrichum truncatum and Phomopsis spp., or Cercospora sojina in soybean seeds. Phytopathology 75:489-492.
297. Lamka, G. 1986. Effects of environmental and genotypic factors on Phomopsis infection of soybean pods. MS thesis, Iowa State University.
298. Laurence, J. A. 1981. Effects of multiple short duration, variable concentration, SO exposures on lesion development by Xanthomonas phaseoli var. sojensis. (Abstr.) Phytopathology 71:234.
299. Laurence, J. A., and Aluisio, A. L. 1981. Effects of sulfur dioxide on expansion of lesions caused by Corynebacterium nebraskense in maize and by Xanthomonas phaseoli var. sojensis in soybean. Phytopathology 71:445-448.
300. Lawn, D. A., and Noel, G. R. 1986. Field interrelationship among Heterodera glycines, Pratylenchus scribneri, and three other nematode species associated with soybean. J. Nematol. 18:98-106.
301. Layton, A. C., Athow, K. L., and Laviolette, F. A. 1986. New physiologic race of Phytophthora megasperma f. sp. glycinea. Plant Disease 70:500-501.
302. Lazarovits, G. 1985. Influence of pyroxyfur seed treatment, inoculum density, and low level cultivar resistance on Phytophthora megasperma f. sp. glycinea of soybean. Can. J. Plant Pathol. 7:370-376.
303. Leach, L. D. 1947. Growth rates of host and pathogen as factors determining the severity of preemergence damping-off. J. Agric. Res. 75:161-179.
304. Leary, J. J., Brigati, D. J., and Ward, D. C. 1983. Rapid and sensitive colorimetric method for visualizing biotin-labeled DNA probes hybridized to DNA or RNA immobilized on nitrocellulose: Bio-blots. Proc. Nat'l. Acad. Sci. U.S.A. 80:4045-4049.
305. Lesney, M. S., and Murakishi, H. H. 1981. Effects of buffer and amendments on infection of soybean callus protoplasts with bean pod mottle virus (BPMV). (Abstr.) Phytopathology 71:236.
306. Lifshitz, C., Simonson, F. M., Scher, J. W., Kloepper, J. W., Rorick-Semple, C., and Zaleska, I. 1986. Effect of rhizobacteria on the severity of phytophthora root rot of soybean. Can. J. Plant Pathol. 8:102-106.
307. Lim, S. M. 1982. A new source of resistance to soybean mosaic virus in a soybean line and its inheritance. (Abstr.) Phytopathology 72:943.
308. Lim, S. M. 1983. Response to Septoria glycines of soybeans nearly isogenic except for seed color. Phytopathology 73:719-722.

309. Lim, S. M., Bernard, R. L., Nickell, C. D., and Gray, L. E. 1982. A new biotype of Peronospora manshurica in soybean disease monitoring plots in Illinois. (Abstr.) Phytopathology 72:974.

310. Lin, M. T., and Hill, J. H. 1983. Bean pod mottle virus: Occurrence in Nebraska and seed transmision in soybeans. Plant Disease 67:230-233.

311. Lister, R. M. 1978. Application of enzyme-linked immunosorbent assay for detecting viruses in soybean seeds and plants. Phytopathology 63:1393-1400.

312. Long, M., Barton-Willis, P., Staskawicz, B. J., Dahlbeck, D., and Keen, N. T. 1985. Further studies on the relationship between glyceollin accumulation and the resistance of soybean leaves to Pseudomonas syringe pv. glycinea. Phytopathology 75:235-239.

313. Long, M., Keen, N. T., Ribeiro, O. K., Leary, J. V., Erwin, D. C., and Zentmeyer, G. A. 1975. Phytophthora megasperma var. sojae: Development of wild-type strains for genetic research. Phytopathology 65:592-597.

314. MacDonald, D. H., Noel, G. R., and Lueschen, W. E. 1980. Soybean cyst nematode, Heterodera glycines, in Minnesota. Plant Disease 64:319-321.

315. Malek, R. B., Melton, T. A., Shurtleff, M. C., Jacobsen, B. J., and Edwards, D. I. Collecting and submitting soil samples for nematode analysis. Rept. Plant Dis. Dep. Plant Pathology, Univ. Ill. 1100: 7 pp.

316. Manandhar, J. B., Hartman, G. L., and Sinclair, J. B. 1986. Colletotrichum destructivum, the anamorph of Glomerella glycines. Phytopathology 76:282-285.

317. Manandhar, J. B., Hartman, G. L., and Sinclair, J. B. 1986. Reaction of soybean germplasm of maturity groups 000 to IV to anthracnose. Biological and Cultural Tests 1:32.

318. Manandhar, J. B., Hartman, G. L., and Sinclair, J. B. 1986. Reaction of soybean germplasm of maturity groups V to X to anthracnose. Biological and Cultural Tests 1:33.

319. Manandhar, J. B., Joshi, S., Bharti, M. P., and Sinclair, J. B. 1984. Soybean seed quality in Nepal influenced by maturity group, planting date and harvest time. (Abstr.) Prog. and Abstrs., World Soybean Research Conf. III:21. Ames, IA.

320. Manandhar, J. B., Kunwar, I. K., Singh, T., Hartman, G. L., and Sinclair, J. B. 1985. Penetration and infection of soybean leaf tissues by Colletotrichum truncatum and Glomerella glycines. Phytopathology 75:704-708.

321. Markham, L. A., and McGuire, J. M. 1981. Location of tobacco ringspot virus in roots of soybean. (Abstr.) Phytopathology 71:894.

322. Martin, K. F., and Walters, H. J. 1982. Infection of soybean by Cercospora kikuchii as affected by dew, temperature, and duration of dew periods. (Abstr.) Phytopathology 72:974.

323. Matthews, S. 1981. Evaluation of techniques for germination and vigor studies. Seed Sci. Technol. 9:543-551.

324. Maury, Y., Bossenec, J. M., Boudazin, G., and Duby, C. 1983. The potential of ELISA in testing soybean seed for soybean mosaic virus. Seed Sci. and Technol. 11:491-503.

325. Maury, Y., Duby, C., Bossenec, J. M., and Boudazin, G. 1985. Group analysis using ELISA. Determination of the level of transmission of soybean mosaic virus in soybean seed. Agronomie 5:405-415.

326. McAlister, D. F., and Krober, O. A. 1951. Translocation of food reserves from soybean cotyledons and their influence on the development of the plant. Pl. Physiol. 26:525-528.

327. McDonald, D. 1986. China spurns soybean shipment. Farm J. 110(8):20-21, May, 1986.

328. McDonald, M. B. 1975. A review and evaluation of seed vigor tests. Proc. Assoc. Off. Seed Anal. 65:109-139.
329. McGawley, E. C., and Chapman, R. A. 1976. Development of concomitant populations of Helicotylenchus pseudorobustus, Paratylenchus projectus, and Criconemoides simile on soybean. (Abstr.) J. Nematol. 8:296.
330. McGawley, E. C., and Chapman, R. A. 1982. Population dynamics of Criconemoides simile on soybean. J. Nematol. 14:572-575.
331. McGawley, E. C., and Chapman, R. A. 1983. Reproduction of Criconemoides simile, Helicotylenchus pseudorobustus, and Paratylenchus projectus on soybean. J. Nematol. 15:87-91.
332. McGee, D. C. 1982. Prevalence of the causal organisms of stem canker and pod and stem blight on soybean pods and seeds. (Abstr.) Phytopathology 72:944.
333. McGee, D. C. 1983. Epidemiology of soybean seed decay by Phomopsis and Diaporthe spp. Seed Sci. and Technol. 11:719-729.
334. McGee, D. C. 1986. Treatment of soybean seeds. In: Seed treatment. Pages 185-200. Ed. Jeffs, K. A. Lavenham, Sufolk, England.
335. McGee, D. C. 1986. Prediction of Phomopsis seed decay by measuring pod infection. Plant Disease 70:329-333.
336. McGee, D. C. 1988. Evaluation of current predictive methods for control of Phomopsis seed decay of soybeans in: T. D. Wyllie and D. H. Scott, eds. Soybean Diseases of the North Central Region. APS Press.
337. McGee, D. C., and Biddle, J. 1985. A comparison between isolates of Diaporthe phaseolorum var. caulivora from soybean seeds in Iowa and stem cankered soybeans. (Abstr.) Phytopathology 75:1332.
338. McGee, D. C., and Brandt, C. L. 1979. Effect of foliar application of benomyl on infection of soybean seeds by Phomopsis in relation to time of inoculation. Plant Dis. Rept. 63:675-677.
339. McGee, D. C., Brandt, C. L., and Burris, J. S. 1980. Seed mycoflora of soybeans relative to fungal interactions, seedling emergence, and carry over of pathogens to subsequent crops. Phytopathology 70:615-617.
340. McGee, D. C., and Nyvall, R. F. 1984. Pod test for foliar fungicides on soybeans. Iowa Coop. Ext. Serv. Pamph. Pm-1136. 3pp.
341. McLaughlin, M. R., Bryant, G. R., Hill, J. H., Benner, H. I., and Durand, D. P. 1980. Isolation of specific antibody to plant viruses by acid sucrose density gradient centrifugation. Phytopathology 70:831-834.
342. Melton, T. A., Jacobsen, B. J., Edwards, D. I., and Noel, G. R. 1985. The soybean cyst nematode problem. Rept. Plant Dis. Dep. Plant Pathology, Univ. of Ill. 501: 14 pp.
343. Melton, T. A., and Jensen, J. O. 1985. Predictive soil sampling and analysis procedures for the soybean cyst nematode. Rept. Plant Dis. Dep. Plant Pathology, Univ. of Ill. 1107: 5 pp.
344. Mendoza, J. B., Mignucci, J. S., Hepperly, P. R., and Riveros, G. 1982. Factors influencing field development of soybean anthracnose in Puerto Rico. (Abstr.) Phytopathology 72:171.
345. Mengistu, A., and Grau, C. R. 1986. Variation in morphological, cultural, and pathological characteristics of Phialophora gregata and Acremonium sp. recovered from soybean in Wisconsin. Plant Disease 70:1005-1009.
346. Mengistu, A., Grau, C. R., and Gritton, E. T. 1986. Comparison of soybean genotypes for resistance to and agronomic performance in the presence of brown stem rot. Plant Disease 70:1095-1098.
347. Meyer, W. A., Sinclair, J. B., and Khare, M. N. 1974. Factors affecting charcoal rot of soybean seedlings. Phytopathology 64:845-849.

348. Mew, T. W., and Kennedy, B. W. 1982. Seasonal variation in populations of pathogenic pseudomonads on soybean leaves. Phytopathology 72:103-105.

349. Milner, M. 1950. Biological processes in stored soybeans. Pages 483-502 in: Soybeans and Soybean Products. Vol. I. K. S. Markely, ed., Interscience Pub., N.Y. 540 pp.

350. Milner, M., and Geddes, W. F. 1946. Grain storage studies. IV. Biological and chemical factors involved in the spontaneous heating of soybeans. Cereal Chem. 23:449-470.

351. Milner, M., and Thompson, J. B. 1954. Physical and chemical consequences of advanced spontaneous heating in stored soybeans. J. Agric. Food Chem. 2:303-309.

352. Minor, H. C., and Brown, E. A. 1985. Soybean resistance to Phomopsis seed decay. Pages 66-69 in: Proceedings of the 15th Soybean Seed Research Conference. Chicago, IL.

353. Minor, H. C., and Brown, E. A. 1986. Characteristics of a soybean genotype resistant to Phomopsis seed decay. Soybean Genetics Newsletter 13:164-165.

354. Minton, N. A., and Cairns, E. J. 1957. Suitability of soybeans var. Ogden and twelve other plants as hosts of the spiral nematode. (Abstr.) Phytopathology 47:313.

355. Morgan-Jones, G. 1985. The Diaporthe / Phomopsis complex of soybeans: Morphology. Pages 1-7 in: Proceedings of the 1984 Conference on the Diaporthe / Phomopsis Disease Complex of Soybean. M. M. Kulik, ed. U.S. Dep. Agric., Agric. Res. Serv. 89 pp.

356. Morgan-Jones, G., and Backman, P. 1984. Characterization of southeastern biotypes of Diaporthe phaseolorum var. caulivora the causal organism of soybean stem canker. (Abstr.) Phytopathology 74:815.

357. Muchovej, J. L., Muchovej, R. M. C., Dhingra, O. D., and Maffia, L. A. 1980. Suppression of anthracnose of soybeans by calcium. Plant Disease 64:1088-1089.

358. Mueller, J. D., Cline, M. N., Sinclair, J. B., and Jacobsen, B. J. 1985. An in vitro test for evaluating efficacy of mycoparsites of sclerotia of Sclerotinia sclerotiorum. Plant Disease 69:584-587.

359. Mueller, J. D., Shortt, B. J., and Sinclair, J. B. 1985. Effects of cropping history, cultivar, and sampling date on the internal fungi of soybean roots. Plant Disease 69:520-523.

360. Mueller, J. D., and Sinclair, J. B. 1986. Occurrence and role of Gliocladium roseum in field-grown soybeans in Illinois. Trans. Brit. Mycol. Soc. 86:677-680.

361. Mulrooney, Robert P. 1986. Soybean disease loss estimate for southern United States. Plant Disease 70:893.

362. Navarro, S., Donahaye, E., and Calderon, M. 1973. Studies on aeration with refrigerated air. II. Chilling of soybeans undergoing spontaneous heating. J. Stored Prod. Res. 9:261-268.

363. Neergaard, P. 1979. Seed Pathology. 2nd ed. London, MacMillan Press. 2 v.

364. Nedrow, B. L., and Harman, G. E. 1980. Salvage of New York soybean seeds following an epiphytotic of seed-borne pathogens associated with delayed harvest. Plant Disease 64:696-698.

365. Nguyen, M. V., Rode, M. W., and Payne, D. W. 1985. Inoculation of seedling cuttings for screening the soybean against races of Phytophthora megasperma f. sp. glycinea. Crop Sci. 25:121-123.

366. Nik, W. Z., and Lim, T. K. 1984. Occurrence and site of infection of Colletotrichum dematium f. sp. truncatum in naturally infected soybean seeds. J. of Plant Prot. in the Tropics. 1:87-91.

367. Nittler, L. W., Harman, G. E., and Nelson, B. 1974. Hila discoloration of Traverse soybean seeds: A problem in cultivar purity analysis and a possible indication of low quality seeds. Proc. Assoc. Off. Seed Anal. 64:115-119.

368. Norton, D. C. 1977. Helicotylenchus pseudorobustus as a pathogen on corn, and its densities on corn and soybean. Iowa State J. of Res. 51:279-285.

369. Norton, D. C. Chair of Committee. 1984. Distribution of plant parasitic nematode species in North America. The Society of Nematologists. 205 pp.

370. Norton, D. C., and Burns, N. 1971. Colonization and sex ratios of Pratylenchus alleni in soybean roots under two soil moisture regimes. J. Nematol. 3:374-377.

371. Norton, D. C., Fredrick, L. R., Ponchillia, P. E., and Nyhan, J. W. 1971. Correlations of nematodes and soil properties in soybean fields. J. Nematol. 3:154-163.

372. Olah, A. F., and Schmitthenner, A. F. 1985. A growth chamber test for measuring Phytophthora root rot tolerance in soybean seedlings. Phytopathology 75:546-548.

373. Olah, A. F., Schmitthenner, A. F., and Walker, A. K. 1985. Glyceollin accumulation in soybean lines tolerant to Phytophthora megasperma f. sp. glycinea. Phytopathology 75:542-546.

374. Oplinger, E. S. 1980. Population and row spacing interaction with soybean cultivars. Pages 47-56 in: Proc. Soybean Seed Res. Conf. 10th Am. Seed Trade Assoc.

375. Orbin, D. P. 1973. Histopathology of soybean roots infected with Helicotylenchus dihystera. J. Nematol. 5:37-40.

376. Orellana, R. G. 1981. Resistance to bud blight in introductions from the germ-plasm of wild soybean. Plant Disease 65:594-595.

377. Orellana, R. G., Reynolds, S. L., Sloger, C., and van Berkum, P. 1983. Specific effects of soybean mosaic virus on total N, Ureide-N, and symbiotic N- fixation activity in Glycine max and G. soja. Phytopathology 73:1156-1160.

378. Ouchi, S. 1983. Induction of resistance and susceptibility. Annu. Rev. of Phytopathol. 21:289-315.

379. Owens, R. A., and Diener, T. O. 1981. Sensitive and rapid diagnosis of potato spindle tuber viroid disease by nucleic acid hybridization. Science 213:670-672.

380. Pacumbaba, R. P., Sapra, V. T., and Prom, L. K. 1984. Effect of two commercial fungicides on incidence of Diaporthe phaseolorum var. caulivora on susceptible soybean cultivars. (Abstr.) Phytopathology 74:827.

381. Palm, E. W. 1984. Evaluation of foliar fungicides on soybeans, 1983. Fungicide and Nematicide Tests 39:141.

382. Park, E. W., and Lim, S. M. 1983. Effects of bacterial blight on soybean yield. (Abstr.) Phytopathology 73:795.

383. Paschal, E. H., II, and Ellis, M. A. 1981. Identification of soybean genotypes with superior seed quality. Pages 71-72 in: F. T. Corbin (ed.), Abstracts, World Soybean Res. Conf. II. Westview Press, Boulder, CO.

384. Pataky, J. K., and Lim, S. M. 1981. Effect of Septoria brown spot on yield components of soybeans. Plant Disease 65:588-590.

385. Pataky, J. K., and Lim, S. M. 1981. Efficacy of benomyl for controlling Septoria brown spot of soybeans. Phytopathology 71:438-442.

386. Paxton, J. D. 1983. Phytophthora root and stem rot of soybean: A case study. Pages 19-30 in: Biochemical Plant Pathology. J. A. Callow ed. John Wiley and Sons, New York.

387. Pearson, C. A. S., Leslie, J. F., and Schwenk, F. W. 1986. Variable chlorate resistance in Macrophomina phaseolina from corn, soybean, and soil. Phytopathology 76:646-649.

388. Pearson, C. A. S., Schwenk, F. W., Crowe, F. J., and Kelley, K. 1984. Colonization of soybean roots by Macrophomina phaseolina. Plant Disease 68:1086-1088.

389. Peterson, D. J., and Edwards, H. H. 1982. Effects of temperature and leaf wetness period on brown spot disease of soybeans. Plant Disease 66:995-998.

390. Pfannenstiel, M. A., Slack, S. A., and Lane, L. C. 1980. Detection of potato spindle tuber viroid in field-grown potatoes by an improved electrophoretic assay. Phytopathology 70:1015-1018.

391. Phillips, D. V. 1983. Development of frogeye leafspot lesions on soybean leaves. (Abstr.) Phytopathology 73:817.

392. Phillips, D. V. 1984. A selective medium for Diaporthe phaseolorum var. caulivora. (Abstr.) Phytopathology 74:815.

393. Phillips, D. V., Hobbs, T. W., Arnett, J. D., and Shokes, F. M. 1985. Influence of applications of thiabendazole on the isolation of Diaporthe biotypes causing southern stem canker of soybean. (Abstr.) Phytopathology 75:1295.

394. Phipps, P. M. 1984. Soybean and peanut seed treatment: New developments and needs. Plant Disease 68:76-77.

395. Phipps, P. M., and Porter, D. M. 1982. Sclerotinia blight of soybean caused by Sclerotinia minor and Sclerotinia sclerotiorum. Plant Disease 66:163-165.

396. Ploetz, R. C., and Shokes, F. M. 1985. Soybean stem canker incited by ascospores and conidia of the fungus causing the disease in the southeastern United States. Plant Disease 69:990-992.

397. Ploetz, R. C., Sprenkel, R. K., and Shokes, F. M. 1986. Current status of soybean stem canker in Florida. Plant Disease 70:600-602.

398. Ploper, L. D. 1987. Influence of soybean genotype on rate of seed maturation and its impact on seedborne fungi. Ph.D. Thesis. Purdue University, West Lafayette, Indiana. 182 pp.

399. Ploper, L. D., and Abney, T. S. 1985. Effect of late season maturation rate on soybean quality (Abstr.) Phytopathology 75:965.

400. Ploper, L. D., Athow, K. L., and Laviolette, F. A. 1985. A new allele at the Rps3 locus for resistance to Phytophthora megasperma f. sp. glycinea in soybean. Phytopathology 75:690-694.

401. Poushinsky, G., and Basu, P. K. 1984. A study of distribution and sampling of soybean plants naturally infected with Pseudomonas syringae pv. glycinea. Phytopathology 74:319-326.

402. Prasartee, C., Tenne, F. D., Ilyas, M. B., Ellis, M. A., and Sinclair, J. B. 1975. Reduction of internally seedborne Diaporthe phaseolorum var. sojae by fungicide sprays. Plant Dis. Rept. 59:20-23.

403. Punithalingam, E. 1983. The nuclei of Macrophomina phaseolina (Tassi) Goid. Nova Hedwigia 38:339-367.

404. Purdy, L. H. 1979. Sclerotinia sclerotiorum: History, diseases, and symptomology, host range, geographic distribution, and impact. Phytopathology 69:875-880.

405. Pyndji, M. M., and Sinclair, J. B. 1987. Soybean seed thermotherapy using heated vegetable oils. Plant Disease 71:213-216.

406. Radke, V. L., and Grau, C. R. 1986. Effects of herbicides on carpogenic germination of _Sclerotinia sclerotiorum_. Plant Disease 70:19-23.

407. Ramstad, P. E., and Geddes, W. F. 1942. The respiration and storage behavior of soybeans. Minn. Agric. Expt. Sta. Tech. Bull. 156. 54 pp.

408. Rebois, R. V., and Golden, A. M. 1985. Pathogenicity and reproduction of _Pratylenchus agilis_ in field microplots of soybeans, corn, tomato, or corn-soybean cropping systems. Plant Disease 69:927-929.

409. Rebois, R. V., and Huettel, R. N. 1986. Population dynamics, root penetration, and feeding behavior of _Pratylenchus agilis_ in monoxenic root cultures of corn, tomato, and soybean. J. Nematol. 18:392-397.

410. Reis, E. M. 1973. [Effect of inoculum concentration of _Colletotrichum dematium_ f. sp. _truncata_ (Schw. von Arx) on reaction of soybean (_Glycine max_ (L.) Merr.) cultivars]. M.S. Thesis, Univ. of Sao Paulo, Brazil (in Portuguese). 43p.

411. Richardson, M. J. 1979. An annotated list of seed-borne diseases. 3rd ed. Phytopathological Papers No. 23. C.A.B.

412. Riggs, R. D., Hamblen, M. L., and Rakes, L. 1981. Infraspecies variation in reactions to hosts in _Heterodera glycines_ populations. J. Nematol. 13:171-179.

413. Robertson, J. A., Morrison, W. H. III, and Burdick, D. 1973. Chemical evaluation of oil from field- and storage-damaged soybeans. J. Amer. Oil Chem. Soc. 50:443-445.

414. Rodriquez-Marcano, A., and Sinclair, J. B. 1981. Variation among isolates of _Colletotrichum dematium_ var. _truncata_ from soybean and three _Colletotrichum_ spp. to benomyl. J. Agr. Univ. Pùerto Rico 66:35-43.

415. Rosenbrock, S., and Wyllie, T. D. 1986. Factors Affecting Populations of _Macrophomina_ in Missouri Soil. M.S. Thesis, Univ. of Missouri-Columbia.

416. Rosenbrock, S., and Wyllie, T. D. 1986. Statistical models for the prediction of microsclerotia of _Macrophomina phaseolina_ in Missouri soils. APS Abstracts of Presentations #250.

417. Ross, J. P. 1975. Effect of overhead irrigation and benomyl sprays on late-season foliar diseases, seed infection, and yields of soybeans. Plant Dis. Rept. 59:809-813.

418. Ross, J. P. 1982. Effect of simulated dew and postinoculation moist periods on infection of soybean by _Septoria glycines_. Phytopathology 72:236-238.

419. Ross, J. P. 1982. Preemptive fungal infection of soybean seed. (Abstr.) Phytopathology 72:974.

420. Ross, J. P. 1986. Registration of eight soybean germplasm lines resistant to seed infection by _Phomopsis_ spp. Crop Sci. 26:210-211.

421. Ross, J. P., Nusbaum, C. J., and Hirschmann, H. 1967. Soybean yield reduction by lesion, stunt, and spiral nematodes. (Abstr.) Phytopathology 57:463-464.

422. Rothrock, C. S., Hobbs, T. W., and Phillips, D. V. 1985. Effects of tillage and cropping system on incidence and severity of southern stem canker of soybean. Phytopathology 75:1156-1159.

423. Roy, K. W. 1976. The mycoflora of soybean reproductive structures. Ph.D. Thesis. Purdue University, West Lafayette, Indiana. 255 pp.

424. Roy, K. W. 1982. Seedling diseases caused in soybean by species of _Colletotrichum_ and _Glomerella_. Phytopathology 72:1093-1096.

425. Roy, K. W., and Abney, T. S. 1976. Purple seed stain of soybeans. Phytopathology 66:1045-1049.

426. Roy, K. W., and Andrews, H. 1984. Resistance of a hardseeded soybean line to seed infection by _Diaporthe phaseolorum_ var. _sojae_. (Abstr.) Phytopathology 74:632.

427. Roy, K. W., and Miller, W. A. 1983. Soybean stem canker incited by isolates of Diaporthe and Phomopsis spp. from cotton in Mississippi. Plant Disease 67:135-137.

428. Rupe, J. C., and Ferriss, R. S. 1982. Influence of environment on infection by Phomopsis sp. of soybean seedlings placed in the field. (Abstr.) Phytopathology 72:1007.

429. Rupe, J. C., and Ferriss, R. S. 1984. A simple model for predicting infection of vegetative soybean tissue by Phomopsis sp. (Abstr.) Phytopathology 74:791.

430. Rupe, J. C., and Ferriss, R. S. 1986. Effects of pod moisture on soybean seed infection by Phomopsis sp. Phytopathology 76:273-277.

431. Rupe, J. C., and Weidmann, G. J. 1986. Pathogenicity of a Fusarium sp. isolated from soybean plants with sudden death syndrome. (Abstr.) Phytopathology 76:1080.

432. Russin, J. S., and Boethel, D. J. 1985. Relationship between threecornered alfalfa hopper damage and pod and stem blight and stem anthracnose diseases of soybean in Louisiana. (Abstr.) Phytopathology 75:1284.

433. Russin, J. S., Boethel, D. J., Berggren, G. T., and Snow, J. P. 1986. Effects of girdling by the threecornered alfalfa hopper on symptom expression of soybean stem canker and associated soybean yields. Plant Disease 7:759-761.

434. Rutherford, F. S., and Ward, E. W. B. 1985. Evidence for genetic control of oospore abortion in Phytophthora megasperma f. sp. glycinea. Can. J. Bot. 63:1671-1673.

435. Rutherford, F. S., Ward, E. W. B., and Buzzell, R. I. 1985. Variation in virulence in successive single-zoospore propagations of Phytophthora megasperma f. sp. glycinea. Phytopathology 75:371-374.

436. Saio, K., Nikunni, I., Ando, Y., Otsuru, M., Terauchi, Y., and Kito, M. 1980. Soybean quality changes during model storage studies. Cereal Chem. 57:77-82.

437. Sanders, R. L. 1985. Soybean leaf blight and seed infection caused by isolates of Cercospora kikuchii of diverse origin. M.S. Thesis. Purdue University, West Lafayette, Indiana. 46 pp.

438. Sar, S., Pettiprez, M., and Albertini, L. 1979. [Effect of different factors particularly nutritional ones (pectin, phenol compounds and lignin) on mycelial expansion and sporulation of Diaporthe phaseolorum (Cke & Ell.) Sacc. var. sojae (Leh.) Wehm. parasitic on soybean (Glycine max (L.) Merr.)]. Phytopathologia Mediterranea 18:10-20 (in French).

439. Sasser, J. N., Barker, K. R., and Nelson, L. A. 1975. Correlations of field populations of nematodes with crop growth responses for determining relative involvement of species. J. Nematol. 7:193-198.

440. Sauer, D. B. 1988. Grain quality and grading standards in: T. D. Wyllie, and D. H. Scott, eds. Soybean Diseases of the North Central Region. APS Press.

441. Schaad, N. W., Azad, H., Peet, R. C., and Panopoulos, N. J. 1986. Cloned phaseolotoxin gene as a hybridization probe for identification of Pseudomonas syringae pv. phaseolicola. (Abstr.) Phytopathology 76:846.

442. Schlub, R. L., and Schmitthenner, A. F. 1978. Effects of soybean seed coat cracks on seed exudation and seedling quality in soil infested with Pythium ultimum. Phytopathology 68:1186-1191.

443. Schmitt, D. P. 1977. Suitability of soybean cultivars to Xiphinema americanum and Macroposthonia ornata. (Abstr.) J. Nematol. 9:284.

444. Schmitt, D. P., and Barker, K. R. 1981. Damage and reproductive potentials of Pratylenchus brachyurus and P penetrans on soybean. J. Nematol. 13:327-332.

445. Schmitt, D. P., and Corbin, F. T. 1981. Interaction of fensulfothion and phorate with preemergence herbicides on soybean parasitic nematodes. J. Nematol. 13:37-41.

446. Schmitt, D. P., and Noel, G. R. 1985. Nematode parasites of soybeans. Pp. 13-39 in: W. R. Nickle, ed. Plant and Insect Nematodes. New York: Marcel Dekker, Inc.

447. Schmitthenner, A. F. 1985. Problems and progress in control of Phytophthora root rot of soybean. Plant Disease 69:362-368.

448. Schmitthenner, A. F. 1988. Phytophthora root rot of soybean in: T. D. Wyllie, and D. H. Scott, eds. Soybean Diseases of the North Central Region. APS Press.

449. Schmitthenner, A. F., and Kmetz, K. T. 1980. Role of Phomopsis sp. in the soybean seed rot problem. Pages 355-366 in: World Soybean Research Conference II: Proceedings. F. T. Corbin, ed. Westview Press, Boulder, Colorado. 897 pp.

450. Schmitthenner, A. F., and Van Doren, Jr., D. M. 1985. Integrated control of Phytophthora root rot of soybean caused by P megasperma f. sp. glycinea, Pp. 263-266 in: Ecology and Management of Soilborne Plant Pathogens. C. A. Parker, A. D. Rovira, K. J. Moore, P. T. W. Wong, and J. F. Killmorgen, eds. American Phytopathological Society, St. Paul, MN.

451. Schneider, R. W., Dhingra, O. D., Nicholson, J. F., and Sinclair, J. B. 1974. Colletotrichum truncatum borne within the seed coat of soybean. Phytopathology 64:154-155.

452. Scott, D. 1984. Chemical control of Phytophthora root rot of soybeans. Pages 74-77 in: Tenth Annual Illinois Crop Protection Workshop Proc. University of Illinois Coop. Ext. Serv., Urbana, IL.

453. Scott, Donald, H. 1986. Soybean sudden death syndrome. Indiana Coop. Ext. Serv., Purdue Pest Management Newsletter, No. 3, April 11, 1986, pp. 7-9.

454. Sebastian, S. A., and Nickell, C. D. 1985. Inheritance of brown stem rot resistance in soybeans Glycine max. J. Hered. 76(3):194-198.

455. Sebastian, S. A., Nickell, C. D., and Gray, L. E. 1986. Relationship between greenhouse and field ratings for brown stem rot reaction in soybean. Crop Sci. 26:665-667.

456. Sherwin, H. S., and Kreitlow, H. 1952. Discoloration of soybean seeds by the fungus Cercospora sojina. Phytopathology 42:568-572.

457. Short, G. E., Wyllie, T. D., and Bristow, P. R. 1980. Survival of Macrophomina phaseolina in soil and in residue of soybean. Phytopathology 70:13-17.

448. Shortt, B. J., Grybauskas, A. P., and Tenne, F. D. 1981. Epidemiology of Phomopsis seed decay of soybean in Illinois. Plant Disease 65:62-64.

459. Shortt, B. J., Jacobsen, B. J., Shurtleff, M. C., and Sinclair, J. B. 1981. Soybean seed quality and fungicide treatment. Rept. on Plant Dis. n.506. Dept. of Plant Pathol., Univ. of Ill., Urbana-Champaign.

460. Shortt, B. J., Mueller, J. D., and Sinclair, J. B. 1980. Chemical and biological soybean seed treatments 1979. Fungicide and Nematicide Tests 35:192-193.

461. Shortt, B. J., Sinclair, J. B., Helm, C. G., Jeffords, M. R., and Kogan, M. 1982. Soybean seed quality losses associated with bean leaf beetles and Alternaria tenuissima. Phytopathology 72:615-618.

462. Shotwell, O. L., Goulden, M. L., Bennett, G. A., Plattner, R. D., and Hesseltine, C. W. 1977. Survey of 1975 wheat and soybeans for aflatoxin, zearalenone, and ochratoxin. J. Assoc. Off. Anal. Chem 60:778-783.

463. Shotwell, O. L., Hesseltine, C. W., Burmeister, H. R., Kwolek, W. R., Shannon, G. M., and Hall, H. H. 1969. Survey for aflatoxin: II. Corn and soybeans. Cereal Chem. 46:454-463.

464. Shotwell, O. L., Vandegraft, E. E., and Hesseltine, C. W. 1978. Aflatoxin formation on sixteen soybean varieties. J. Assoc. Off. Anal. Chem. 61:574-577.

465. Shurtleff, M. C., Jacobsen, B. J., and Sinclair, J. B. 1980. Pod and Stem Blight of Soybean. Rept. on Plant Dis., No. 509 (revised), Dept. Plant Path., Univ. Ill., Urbana-Champaign.

466. Siddiqui, M. R., Mathur, S. B., and Neergaard, P. 1983. Longevity and pathogenicity of Colletotrichum spp. in seed stored at 5 C. Seed Sci. and Technol. 11:353-361.

467. Sim IV, T., and Todd, T. C. 1986. First field observation of the cyst nematode in Kansas. Plant Disease 70:603.

468. Simon, E. W. 1984. Early events in seed germination. Pages 77-115 in: Seed Physiology. Vol. 2: Germination and Reserve Mobilization. D. R. Murray, ed., Acad. Pr., N.Y. 295 pp.

469. Sinclair, J. B. 1982. (ed.) Compendium of Soybean Diseases. 2nd ed. American Phytopathological Society, St. Paul, MN. 104 pp.

470. Sinclair, J. B. (ed.) 1983. Compendium of Soybean Diseases. The American Phytopathological Society, St. Paul, MN.

471. Sinclair, J. B. 1988. Anthracnose of soybeans in: T. D. Wyllie, and D. H. Scott, eds. Soybean Diseases of the North Central Region. APS Press.

472. Sinclair, J. B. 1988. Diaporthe/Phomopsis complex of soybeans in: T. D. Wyllie, and D. H. Scott, eds. Soybean Diseases of the North Central Region. APS Press.

473. Singh, T., and Sinclair, J. B. 1986. Further studies on the colonisation of soybean seeds by Cercospora kikuchi and Phomopsis sp. Seed Sci. and Technol. 14:71-77.

474. Smith, E. F. 1987. Epidemiology of Diaporthe phaseolorum var. caulivora in southern soybean. M.S. Thesis, Dept. of Plant Pathology, Auburn University, AL.

475. Smith, E. F., and Backman, P. A. 1988. Soybean Stem Canker: An Overview in: T. D. Wyllie, and D. H. Scott, eds. Soybean Diseases of the North Central Region. APS Press.

476. Smith, W. H. 1969. Comparison of mycelial and sclerotial inoculum of Macrophomina phaseoli in the mortality of pine seedlings under varying soil conditions. Phytopathology 59:379-382.

477. Sonku, Y., Simanuki, T., and Charcher, M. J. D. A. 1982. Identification of species of Stylosanthes anthracnoses in Brazil and their physiologic specialization. Fitopatologia Brasieleira 7:483.

478. Sortland, M. E., and MacDonald, D. H. 1982. Development of a Minnesota population of Heterodera glycines Race 5 at four soil temperatures. (Abstr.) J. Nematol. 14:471.

479. Specht, J. E., Williams, J. H., and Pearson, D. R. 1985. Near-isogenic analyses of soybean pubescence genes. Crop Sci. 25:92-96.

480. Spilker, D. A., Schmitthenner, A. F., and Ellet, C. W. 1981. Effects of humidity, temperature, fertility, and cultivar on the reduction of soybean seed quality by Phomopsis sp. Phytopathology 71:1027-1029.

481. Steadman, J. R. 1983. White mold - A serious yield-limiting disease of bean. Plant Disease 67:346-350.

482. Stuckey, R. E., Jacques, R. M., TeKrony, D. M., and Egli, D. B. 1981. Foliar fungicides can improve seed quality. Kentucky Seed Improv. Assoc., Lexington, KY.

483. Stuckey, R. E., Moore, W. F., and Wrather, J. A. 1984. Predictive systems for scheduling foliar fungicides on soybeans. Plant Disease 68:743-744.

484. Stuckey, R. E., Wilcox, W. F., and Clinton, W. 1983. Evaluation of foliar fungicides for disease control and yield of soybeans, 1982. Fungicide and Nematicide Tests 38:82.

485. Surico, G., Kennedy, B. W., and Ercolani, G. L. 1981. Multiplication of Pseudomonas syringae on soybean primary leaves exposed to aerosolized inoculum. Phytopathology 71:532-536.

486. Sutherland, E. D., and Lockwood, J. L. 1984. Hyperparasitism of oospores by some Peronosporales by Actinoplanes missouriensis and Humicola fuscoatra and other Actinomycetes and fungi. Can. J. Plant Pathol. 6:139-145.

487. Sutton, B. C. 1980. The Coelomycetes. Commonwealth Mycological Institute, Kew, Surrey, England. 696p.

488. Sutton, D. C., and Deverall, B. J. 1983. Studies on infection of bean (Phaseolus vulgaris) and soybean (Glycine max) by ascospores of Sclerotinia sclerotiorum. Plant Pathology 32:251-261.

489. Sutton, D. C., and Deverall, B. J. 1984. Phytoalexin accumulation during infection of bean and soybean by ascospores and mycelium of Sclerotinia sclerotiorum Plant Pathology 33:377-383.

490. Sutula, C., Gillett, J. M., Morrissey, S. M., and Ramsdell, D. C. 1986. Interpreting ELISA data and establishing the positive-negative threshold. Plant Disease 70:722-726.

491. Tachibana, H. 1982. Prescribed resistant cultivars for control of brown stem rot of soybean and similar plant diseases and resistant gene management. Plant Disease 66:271-274.

492. Tachibana, H. 1987. Brown stem rot of soybeans: 1948-2000. Proc. 16th Soybean Seed Res. Conf., Amer. Seed Trade Assoc., Chicago, IL, Dec. 9-10, 1986, pp. 84-94.

493. Tachibana, H. 1988. Use and management of resistance for control of brown stem rot of soybeans in: T. D. Wyllie, and D. H. Scott, eds. Soybean Diseases in the North Central Region. APS Press.

494. Tachibana, H., and Booth, G. D. 1979. An efficient and objective survey method for brown stem rot of soybeans. Plant Dis. Rept. 63:539-541.

495. Tachibana, H., and Card, L. C. 1972. Brown stem rot resistance and its modification by soybean mosiac virus in soybean. Phytopathology 62:1314-1317.

496. Tachibana, H., and Card, L. C. 1979. Field evaluation of soybeans resistant to brown stem rot. Plant Dis. Rept. 63:1042-1045.

497. Tachibana, H., Card, L. C., Bahrenfus, J. B., and Fehr, W. R. 1980. Registration of BSR 301 soybean. Crop Sci. 20:414-415.

498. Tachibana, H., Hatfield, J. D., and Higley, P. M. 1983. Phytophthora megasperma var. glycinea suppressive soil of soybean. (Abstr.) Phytopathology 78:823.

499. Tachibana, H., and Van Diest, A. 1983. Association of Phytophthora root rot of soybean with conservation tillage. (Abstr.) Phytopathology 73:844.

500. Taylor, A. L., and Sasser, J. N. 1978. Biology, identification, and control of root-knot nematodes (Meloidogyne species). Raleigh, North Carolina: North Carolina State University Graphics. 111 pp.

501. Taylor, D. P. 1961. Biology and host-parasite relationships of the spiral nematode, Helicotylenchus microlobus. Proceedings of the Helminthol. Soc. of Washington 28:60-66.

502. Taylor, D. P., and Wyllie, T. D. 1959. Interrelationship of root knot nematodes and Rhizoctonia solani on soybean emergence. (Abstr.) Phytopathology 49:552.

503. Taylor, P. L., Ferris, J. M., and Ferris, V. R. 1972. Technique for quantifying injury to seedling soybeans by Pratylenchus penetrans without sacrificing the plant. J. Nematol. 4:68-69.

504. TeKrony, D. M., Egli, D. B., Balles, J., Thomes, L., and Stuckey, R. E. 1984. Effect of date of harvest maturity on soybean seed quality and Phomopsis sp. seed infection. Crop Sci. 24:189-193.
505. TeKrony, D. M., Egli, D. B., Stuckey, R. E., and Balles, J. 1983. Relationship between weather and soybean seed infection by Phomopsis sp. Phytopathology 73:914-918.
506. TeKrony, D. M., Egli, D. B., Stuckey, R. E., and Loeffler, T. M. 1985. Effect of benomyl applications on soybean seedborne fungi, seed germination and yield. Plant Disease 69:763-765.
507. TeKrony, D. M., Stuckey, R. E., Egli, D. B., and Tomes, I. 1985. Effectiveness of a point system for scheduling foliar fungicides in soybean seed fields. Plant Disease 69:962-965.
508. Tenne, F. D., Foor, S. R., and Sinclair, J. B. 1977. Association of Bacillus subtilis with soybean seeds. Seed Sci. Technol. 5:763-769.
509. Teo, B. K., and Morrall, R. A. A. 1985. Influence of matric potentials on carpogenic germination of sclerotia of Sclerotinia sclerotiorum. I. Development of an inclined box technique to observe apothecium production. Can. J. Plant Pathol. 7:359-364.
510. Teng, P. S., Close, R. C., and Blackie, M. J. 1979. Comparion of models for estimating yield loss caused by leaf rust on Zephyr barley in New England. Phytopathology 69:1239-1244.
511. Thapliyal, P. N., Cubeta, M. A., and Sinclair, J. B. 1982. Control of prickly sida plants with Colletotrichum gloeosporioides in Illinois. (Abstr.) Phytopathology 72:1140.
512. Thomason, I. J., Rich, J. R., and O'Melia, F. C. 1976. Pathology and histopathology of Pratylenchus scribneri infecting snap bean and lima bean. J. Nematol. 8:347-352.
513. Thomison, P. R. 1985. Factors affecting the severity of Phomopsis seed decay in soybeans. Pages 495-502 in: Proceedings of the World Soybean Research Conf. III. R. Shibles, ed. Westview Press, Boulder, Colorado. 1262 pp.
514. Thomison, P. R., and Kenworthy, W. J. 1986. Phomopsis seed infection and seed quality in soybean isolines differing in growth habit and maturity. (Abstr.) Phytopathology 76:566.
515. Thompson, A. H., and Van der Westhuizen, G. C. A. 1979. Sclerotinia sclerotiorum (Lib.) De Bary on soybean in South Africa. Phytophylactica 11:145-148.
516. Tiffany, L. H., and Gilman, J. C. 1954. Species of Colletotrichum from legumes. Phytopathology 46:52-75.
517. Tooley, P. W., Bergstrom, G. C., and Wright, M. J. 1984. Widely virulent isolates of Phytophthora megasperma f. sp. glycinea causing root and stem rot of soybeans in New York. Plant Disease 68:726-727.
518. Tooley, P. W., and Grau, C. R. 1984. Field characterization of rate-reducing resistance of Phytophthora megasperma f. sp. glycinea in soybean. Phytopathology 74:1201-1208.
519. Tooley, P. W., and Grau, C. R. 1984. The relationship between rate-reducing resistance to Phytophthora megasperma f. sp. glycinea and yield of soybean. Phytopathology 74:1209-1216.
520. Tooley, P. W., and Grau, C. R. 1986. Microplot comparison of rate-reducing and race-specific resistance to Phytophthora megasperma f. sp. glycinea in soybean. Phytopathology 76:554-557.
521. Tooley, P. W., Grau, C. R., and Stough, M. C. 1982. Races of Phytophthora megasperma f. sp. glycinea in Wisconsin. Plant Disease 66:472-475.

522. Turner, J. T., Jr. 1982. Effects of fungicides and harvest times on soybean seed quality. (Abstr.) Phytopathology 72:362.
523. U.S. Dept. of Agriculture. 1980. Soybeans. Pages 6.1-6.28 in: Grain Inspection Handbook, Book 2. Federal Grain Inspection Service, Washington, D. C.
524. Valverde, R. A., Dodds, J. A., and Heick, J. A. 1986. Double-stranded ribonucleic acid from plants infected with viruses having elongated particles and undivided genomes. Phytopathology 76:459-465.
525. Van Regenmortel, M. H. V. 1982. Serology and Immunochemistry of Plant Viruses. Academic Press, New York. 302p.
526. Vaughn, D. A., Bernard, R. L., and Sinclair, J. B. 1985. Development of soybean seedcoat structures: Relevance to field seed pathogen infection and breeding for resistance. (Abstr.) Phytopathology 75:1331.
527. von Arx, J. A. 1957. Die arten der galtun Colletotrichum. Phytopathol. Z. 29:413-468.
528. von Qualen, R. H. 1987. The influence of rotation and tillage on late season soybean diseases. Ph.D. Thesis. Purdue University, West Lafayette, Indiana. 143 pp.
529. Walker, A. K., and Schmitthenner, A. F. 1984. Comparison of field and greenhouse evaluations for tolerance to Phytophthora root rot in soybean. Crop Sci. 24:487-489.
530. Walker, A. K., and Schmitthenner, A. F. 1984. Heritability of tolerance to Phytophthora rot in soybean. Crop Sci. 24:490-491.
531. Walker, A. K., and Schmitthenner, A. F. 1984. Recurrent selection for tolerance to Phytophthora rot in soybean. Crop Sci. 24:495-497.
532. Wall, M. T., McGee, D. C., and Burris, J. S. 1983. Emergence and yield of fungicide-treated soybean seed differing in quality. Agron. J. 75:969-973.
533. Wallace, H. A. H. 1960. Factors affecting subsequent germination of cereal seeds sown in soils of subgermination moisture content. Can. J. Bot. 38:287-306.
534. Wallen, V. R., and Seaman, W. L. 1963. Seed infection of soybean by Diaporthe phaseolorum and its influence on host development. Can. J. Bot. 41:13-21.
535. Walters, H. J. 1980. Soybean leaf blight caused by Cercospora kikuchii. Plant Disease 64:961-962.
536. Walters, H. J. 1983. Evaluation of fungicides for control of foliar, pod and stem diseases of soybean, 1982. Fungicide and Nematicide Tests 38:82.
537. Walters, H. J., and Caviness, C. E. 1968. Response of Phytophthora resistant and susceptible soybean varieties to 2,4-DB. Plant Dis. Rept. 52:355-357.
538. Ward, E. W., B., Stossel, P., and Lazarowits, G. 1981. Similarities between age-related and race-specific resistance of soybean hypocotyls to Phytophthora megasperma var. sojae. Phytopathology 71:504-508.
539. Watanabe, T. R., Smith, R. S., and Snyder, W. C. 1967. Populations of microsclerotia of the soil-borne pathogen, Macrophomina phaseoli, in relation to stem blight of bean. (Abstr.) Phytopathology 57:1010.
540. Weaver, D. B., Cosper, B. H., Backman, P. A., and Crawford, M. A. 1984. Cultivar resistance to field infestations of soybean stem canker. Plant Disease 68:877-879.
541. Weiss, F. 1946. Check list revision. Plant Dis. Rept. 30:130-137.
542. Welch, A. W., and Gilman, J. C. 1948. Hetero- and homothallic types of Diaporthe of soybeans. Phytopathology 38:628-637.
543. Whitney, N. G. 1983. Effect of foliar fungicides on control of anthracnose on four soybean varieties, 1982. Fungicide and Nematicide Tests 38:83.

544. Whitney, N. G. 1984. Evaluation of foliar fungicides for disease control and yield of soybeans, 1983. Fungicide and Nematicide Tests 39:143.
545. Whitney, N. G. 1985. Effect of fungicides on disease control and yield of soybeans, 1984. Fungicide and Nematicide Tests 40:157.
546. Whitney, N. G. 1986. Evaluation of foliar fungicides of soybeans, 1985. Fungicide and Nematicide Tests 41:118.
547. Whitney, N. G., and Bowers, G. R., Jr. 1985. Stem canker of soybean in Texas. Plant Disease 69:361.
548. Wilcox, J. R., and Abney, T. S. 1973. Effects of _Cercospora kikuchii_ on soybeans. Phytopathology 63:796-797.
549. Wilcox, J. R., Abney, T. S., and Frankenberger, E. M. 1985. Relationship between seedborne soybean fungi and altered photoperiod. Phytopathology 75:797-800.
550. Wilcox, J. R., Laviolette, F. A., and Athow, K. L. 1974. Deterioration of soybean seed quality associated with delayed harvest. Plant Dis. Rept. 58:130-133.
551. Wilcox, W. F., Stuckey, R. E., and Bachi, P. R. 1984. Effect of foliar fungicides on disease control and yield of soybeans, 1983. Fungicide and Nematicide Tests 39:143.
552. Williams, J. R., and Stelfox, D. 1980. Influence of farming practices in Alberta on germination and apothecium production of sclerotia of _Sclerotinia sclerotiorum_. Can. J. Plant Pathol. 2:169-172.
553. Willis, W. G. 1983. New developments in cereal and soybean seed treatment fungicides. Plant Disease 67:257-258.
554. Windham, M. T., and Ross, J. P. 1981. Plant height effects on disease incidence and symptom severity of bean pod mottle virus infection of soybeans. (Abstr.) Phytopathology 71:913.
555. Wong, C. F. J., Nik, W. Z., and Lim, T. K. 1983. Studies of _Colletotrichum dematium_ f. sp. _truncatum_ on soybean. Pertanika 6:28-33.
556. Woodstock, L. W. 1969. Seedling growth as a measure of seed vigor. Proc. Int. Seed Test Assoc. 34:273-280.
557. Woodstock, L. W. 1973. Physiological and biochemical tests for seed vigor. Seed Sci. Technol. 1:127-157.
558. Wright, F. E. 1983. Evaluation of foliar applied fungicides on soybean, 1982. Fungicide and Nematicide Tests 38:84.
559. Wyllie, T. D. 1988. Charcoal rot of soybeans-current status in: T. D. Wyllie, and D. H. Scott, eds. Soybean Diseases of the North Central Region. APS Press.
560. Wyllie, T. D., Gangopadhyay, S., Teague, W. R., and Blanchar, R. W. 1984. Germination and production of _Macrophomina phaseolina_ microsclerotia as affected by oxygen and carbon dioxide concentration. Plant and Soil 81:195-201.
561. Wyllie, T. D., and Taylor, D. P. 1960. Phytophthora root rot of soybeans as affected by soil temperature and _Meloidogyne hapla_. Plant Dis. Rept. 44:543-545.
562. Yaklich, R. W. 1985. Effects of aging on soluble oligosaccharide content in soybean seeds. Crop Sci. 25:701-704.
563. Yaklich, R. W., Vigil, E. L., and Wergin, W. P. 1986. Pore development and seed coat permeability in soybean. Crop Sci. 26:616-624.
564. Yopp, J., Krishnamani, M. R. S., Bozzola, J., Richardson, J., Myers, O., and Klubek, B. 1986. Presumptive role of a pathovar of _Xanthomonas_ in sudden death syndrome of soybean. Microbios Letters 32:75-79.

565. Yorinori, J. T., and Homechin, M. 1985. Sclerotinia stem rot of soybeans, its importance and research in Brazil. pp. 582-588 in: R. Shibles (ed.) Proc. World Soybean Res. Conf. III. Westview Press, Boulder, CO.
566. Yorinori, J. T., and Sinclair, J. B. 1983. Harvest and assay methods for seedborne fungi in soybeans and their pathogenicity. Intl. J. Trop. Plant Dis. 1:53-59.
567. Yoshikawa, M., and Masago, H. 1982. Biochemical mechanism of glyceollin accumulation in soybean. Pages 265-280 in: Plant Infection: The Physiological and Biochemical Basis. Y. Asada, ed. Japan Soc. Sci. Press, Tokyo.
568. Ziegler, E., and Pontzen, R. 1982. Specific inhibition of glucan-elicited glyceollin accumulation in soybeans by an extracellular mannan-glycoprotein of *Phytophthora megasperma* f. sp. *glycinea*. Physiol. Plant Pathol. 20:321-331.
569. Zimmerman, S. 1986. Effect of phosphorus and potassium on the growth and behavior of *Macrophomina phaseolina* (Tassi) Goid. in axenic culture and amended natural soil. M.S. Thesis, Univ. of Missouri-Columbia. 126 p.
570. Zinnen, T. M., and Sinclair, J. B. 1982. Thermotherapy of soybean seeds to control seedborne fungi. Phytopathology 72:831-834.
571. Zirakparvar, M. E. 1980. Host range of *Pratylenchus hexincisus* and its pathogenicity on corn, soybean, and tomato. Phytopathology 70:749-753.
572. Zirakparvar, M. E. 1982. Susceptibility of soybean cultivars and lines to *Pratylenchus hexincisus*. J. Nematol. 14:217-220.